Magnetic Field Assisted Finishing

Advanced Manufacturing Techniques
Series Editor: Arshad Noor Siddiquee

The objective of this series is to cover newer manufacturing techniques such as nano-structured materials, functional and functionally graded materials, micro-electro manufacturing systems, additive manufacturing, and single molecule sensing systems. Aimed at senior undergraduate students, graduate students and professionals, the proposed series will focus on the advanced manufacturing techniques and its applications. Some of the emerging application areas include aerospace engineering, biomedical engineering, defense, and marine construction.

Magnetic Field Assisted Finishing: Methods, Applications and Process Automation
Dilshad Ahmad Khan, Zafar Alam and Faiz Iqbal

For more information about this series, please visit: https://www.routledge.com/Advanced-Manufacturing-Techniques/book-series/AMT

Magnetic Field Assisted Finishing

Methods, Applications and Process Automation

Edited by

Dilshad Ahmad Khan
Zafar Alam
Faiz Iqbal

CRC Press is an imprint of the
Taylor & Francis Group, an **informa** business

First Edition published 2022
by CRC Press
6000 Broken Sound Parkway NW, Suite 300, Boca Raton, FL 33487-2742

and by CRC Press
2 Park Square, Milton Park, Abingdon, Oxon, OX14 4RN

First Edition published by CRC Press 2022

CRC Press is an imprint of Taylor & Francis Group, LLC

ISBN: 978-0-367-75438-9 (hbk)
ISBN: 978-1-032-13337-9 (pbk)
ISBN: 978-1-003-22877-6 (ebk)

DOI: 10.1201/9781003228776

Typeset in Times
by KnowledgeWorks Global Ltd.

Contents

Preface

This disquisition on *Magnetic Field Assisted Finishing: Methods, Applications and Process Automation* briefly discusses traditional finishing processes and, in detail, covers all recently developed processes in the category of magnetic field assisted finishing and the advancement associated thereof. Recent engineered products need precise finishes due to their enhanced surface characteristics in high-tech applications such as aerospace, biomedical, laser, optical, etc. The processes in this category are deterministic, in a sense, so that the finishing results can be replicated at any point of time for the required level of surface finishing. Conventional finishing processes possess difficulties in achieving a very high level of surface finish due to the involvement of a series of finishing operations that depend primarily on the skill of the operator. Other than achieving a high level of surface finish, shape limitations, uniform finish, and selective finishing for a region on the workpiece are also a challenge for conventional finishing processes. Magnetic field assisted finishing processes utilize magnetic finishing media either in dry or fluid form whose stiffness or, in turn, finishing intensity can be controlled by external means, making these processes deterministic. Due to their flexible finishing medium, complex and freeform surfaces can be finished precisely. Also, because of the deterministic nature of the processes, uniform finishing and selective finishing can be achieved. The primary processes in this category are magnetic abrasive finishing (MAF), magnetorheological finishing (MRF), and magnetorheological abrasive flow finishing (MRAFF). Every finishing process-related chapter also explains hybrid finishing processes that have been developed to overcome the limitations of individual processes and combine the benefits of all processes involved. Precise and controlled finishing cannot be achieved without a proper automation of the processes. Therefore, this book has a dedicated chapter for the process automation of magnetic field assisted finishing processes.

The book is organized to impart in-depth knowledge of different primary magnetic field assisted finishing processes and hybrid magnetic field assisted finishing processes, their equipment details, and specific high-end applications. It is organized in five chapters: (1) Introduction; (2) Magnetic Abrasive Finishing; (3) Magnetorheological Finishing; (4) Magnetorheological Abrasive Flow Finishing; and (5) Process Automation of Magnetic Field Assisted Finishing. The variants of these processes include magnetic float polishing (MFP), magnetorheological jet finishing (MRJF), magnetorheological abrasive honing (MRAH), and rotational magnetorheological abrasive flow finishing (R-MRAFF). As some benefits and drawbacks are always associated with any process, some hybrid processes have also been developed in the recent past by combining one or more processes with a particular magnetic field assisted finishing process to overcome the negative aspects and gain the benefits of positive aspects of all processes involved. Hybridization enhances the process capability and, hence, the application areas. Therefore, considering the hybridization of the finishing processes with magnetic field assisted finishing processes evolves a new category of hybrid magnetic field assisted finishing processes, including electrolytic magnetic abrasive finishing (EMAF) and vibration assisted magnetic abrasive finishing.

The chapters cover all aspects of the process, including the working principles, physics involved behind the finishing, variants of underlying processes, and details of their equipment, limitations, and process parameters. Every chapter is well supported by related diagrams and photos for better realization.

The different process parameters have been explained by considering their effects on the finishing output. The bibliography at the end of each chapter complements the chapter explanations considering recent literature archives. The contents of this book are proposed to cater to the needs of academicians, engineers, researchers, and practitioners. This book is an equal combined effort of all three of us, and we would like to express our sincere thanks to Prof. Arshad Noor Siddiqui (Department of Mechanical Engineering, Jamia Millia Islamia, India) for giving us the great opportunity of writing this book and for showing his faith in us. We express our gratitude towards Prof. Sunil Jha (Department of Mechanical Engineering, IIT Delhi, India) for continuous motivation, encouragement, and support. We would also like to thank Taylor & Francis Group (CRC Press) for showing their interest in publishing a book on such an advanced topic.

We will happily acknowledge suggestions from the readers of this book to improve the content and will try to incorporate the appropriate suggestions in the next edition.

Dilshad Ahmad Khan
Zafar Alam
Faiz Iqbal

About the Authors

Dr. Dilshad Ahmad Khan is an Assistant Professor at the Department of Mechanical Engineering, NIT Hamirpur (Himachal Pradesh). He is a prominent academician and researcher. He is a chartered engineer (CEng) from the Institution of Mechanical Engineers, London. He received his PhD in manufacturing from IIT Delhi in 2018 and Master of Technology from Aligarh Muslim University, Aligarh, India, in 2010. He received his Bachelor of Engineering in Mechanical Engineering from Dr. B.R. Ambedkar University, Agra (formerly Agra University), India, in 2006. He has published a number of research articles in international journals and conferences. He has published several book chapters on various topics. He has filed many Indian patents. He has received various national and international awards for his research and innovations. His research interests include advanced finishing/polishing processes, non-conventional machining, mechatronic systems, and industrial automation.

Dr. Zafar Alam is an Assistant Professor at the Department of Mechanical Engineering, Indian Institute of Technology (Indian School of Mines) Dhanbad, Jharkhand, India. Prior to joining IIT (ISM) Dhanbad, he served as Assistant Professor of Mechanical Engineering at Zakir Husain College of Engineering and Technology, Aligarh Muslim University, India, from August 2018 to June 2020. He received his B. Tech in Mechanical Engineering from Jamia Millia Islamia University, New Delhi, India, in 2012. He has received his M. Tech and PhD in Production Engineering from the Indian Institute of Technology Delhi, India, in 2014 and 2019, respectively. As an academician and researcher, he has published numerous research papers in peer-reviewed international journals and conferences. He also has to his credit three Indian patents and has received two international and three national awards, including the critically acclaimed GYTI (Gandhian Young Technological Innovation) Award for his contributions in the field of research and innovation. His research interests include, but are not limited to, advanced finishing/polishing processes, non-conventional machining, industrial automation, and motion control.

Dr. Faiz Iqbal received his B. Tech in Mechanical and Automation Engineering from Maharshi Dayanand University, Rohtak, Haryana, India, in 2011. He received his M. Tech in Mechatronics Engineering from Amity University, Noida, Uttar Pradesh, India, in 2013, and his PhD in Manufacturing Automation from the Indian Institute of Technology Delhi in 2019. He is currently working as a post-doctoral researcher in the Institute of Integrated Micro and Nano Systems, School of Engineering, The University of Edinburgh, Scotland. As an academician and researcher, he has to his credit several research papers in peer-reviewed international journals and conferences. He also holds two Indian patents and has received two international and three national awards for his contribution in the field of research and innovation.

He secured funding for a COVID-19 project, which he successfully delivered and was nominated for the Scottish Knowledge Exchange award in the COVID-19 collaborative response category. His research interests include, but are not limited to, manufacturing automation, advanced manufacturing, industrial automation, surface metrology, non-conventional machining, machining processes and analysis, and mechatronic systems.

1 Introduction

1.1 SURFACE AND ITS CHARACTERISTICS

A surface of an object can be defined as the outermost boundaries, which separate the bulk volume of the object from the outer environment. All the properties such as physical and mechanical are limited up to the surface of the object under consideration. From the design and manufacturing point of view, the performance of the product is affected by many factors among which surface quality is one of the predominant factors. Sometimes surface quality becomes so important that it could be a parameter for selecting or rejecting the component. The performance of the product is counted in terms of various parameters like lifespan of the product, assembly fit, satisfactory working of the product, and aesthetic appeal, etc. These parameters are greatly affected by the surface quality.

A surface having microcracks or residual stresses becomes more prone to complete failure of the component. The importance of surface quality can be understood from the fact that the surface of an object that does not meet the standards sometime becomes a cause of rejection. The characteristics that define a surface quality are called surface characteristics and predominantly depend on manufacturing processes used and the environmental condition at the time of production. For example, the surface produced by milling will show different surface characteristics than the surface produced by a grinding operation. Likewise, the surface produced in dry machining conditions is different from the surface produced under wet machining conditions. The major surface characteristics, which describe the quality of the surface of an object, are the surface texture, surface integrity, and dimensional and geometrical tolerances. The elements of surface texture are defined as surface roughness, waviness, lay, and flaws [1]. These elements are discussed in the following sections.

1.1.1 Surface Roughness

All the manufacturing processes leave some degree of unevenness in their processing; surface roughness is one of them. Surface roughness can be defined as the deviation of finely spaced irregularities on the surface from the nominal surface of the component. Surface roughness is dependent on material characteristics, machining conditions, and the process used.

These deviations are expressed in different orders of international standards. Very finely distributed deviations of lower frequency are referred to as roughness.

First-order deviation is related to form error of the surface, which includes flatness error, circularity error, etc. Second-order deviation is concerned with the waviness on the form error. The first- and second-order deviations are the result of inaccurate

DOI: 10.1201/9781003228776-1

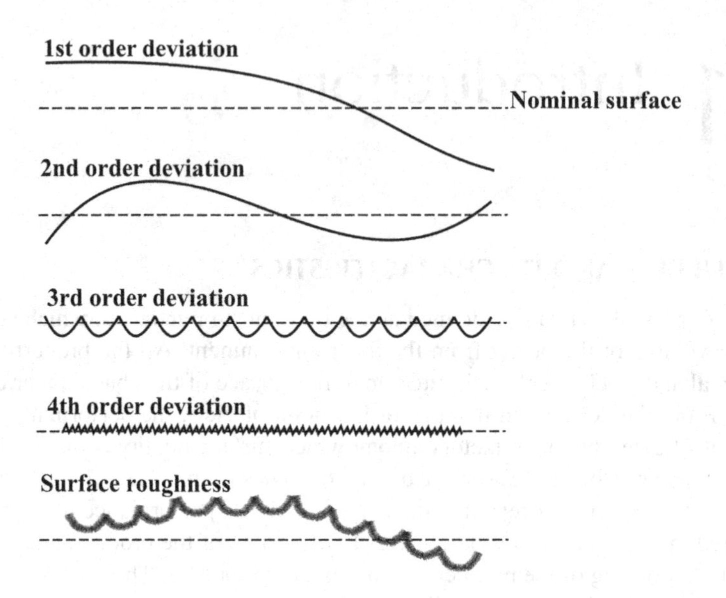

FIGURE 1.1 Forms of surface deviations. (From ref. [2].)

machine tools and cutting tools, erroneous setups and clamping, inhomogeneous workpiece material, deformation of the workpiece, and vibrations. Third- and fourth-order deviations are the results of the condition and shape of the cutting edge, cutting process mechanism, and chip formation. These deviations are related to regular interval grooves, dilapidations, and surface cracks.

The deviations of fifth-order and sixth-order are mainly concerned with the internal structure of the material of the workpiece. If the workpiece material is having internal flaws, such as oxidation, slip, diffusion, residual stresses, etc., then the fifth- and sixth-order deviations appear.

The surface roughness profile is the superimposition of all these different types of deviations as shown in Figure 1.1 [2]. Lesser value of surface roughness is an indication of better surface finishing and vice versa.

1.1.2 Waviness

The deviations of larger frequencies are called waviness. The repetition of the surface is in the form of waves of larger spacing. The waviness occurs due to workpiece deflection, tool chatters, heavy vibrations of machine tool, heat treatment, material inclusions, etc. Surface roughness is generally superimposed in the waviness; hence, the surface roughness is evaluated on a small length of the workpiece, while waviness is evaluated on a longer length.

1.1.3 Lays

Lays are basically the directional marks of a cutting tool on the surface. A specific manufacturing process renders some specific patterns of surface texture based on which the manufacturing process employed can be recognized.

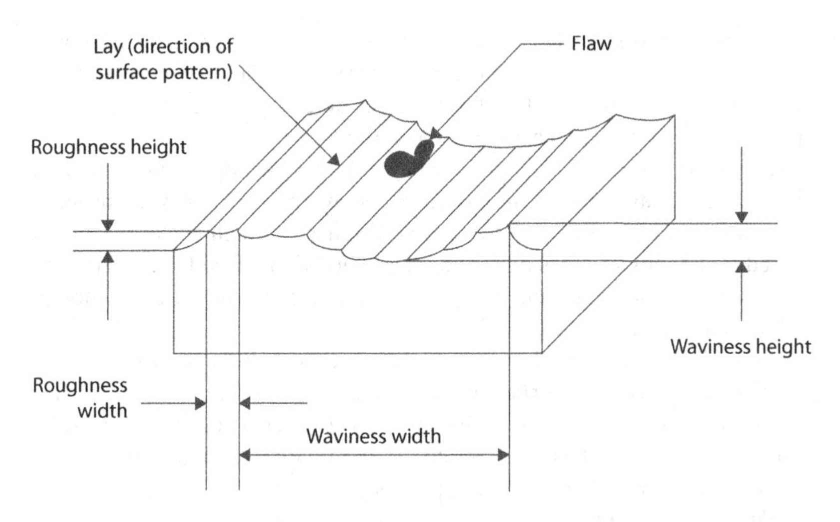

FIGURE 1.2 Surface characteristics. (From ref. [1].)

1.1.4 Flaws

Flaws can be defined as the faults or irregularities. These irregularities sometimes occur on the surface of the component or the part produced. Flaws are basically defects that are not desirable but for various reasons appear. The flaws are not the characteristics of the manufacturing process, but they are considered faults. There is a big list of flaws, which include impurity, cracks, inclusions, scratches, holes, etc. Figure 1.2 shows waviness, lay, and flaw in a surface schematically.

1.2 OBJECTIVES OF GOOD SURFACE FINISH

Sometimes in engineering applications the surfaces come in direct contact and impart some specific characteristics to the assembly. In different engineering applications, different grades of surface finish are required. In some specific applications, the rougher surfaces are suitable and at others a very fine surface finish is the prime requirement. There are various technological as well as commercial reasons associated with surfaces and their grade of finishing. The following are technological and commercial aspects or objectives for which surface finishing operations are carried out:

a. *Enhanced fatigue strength and fatigue life:* Surface with any surface defect is considered poor from a technological point of view. Surface defects act as the stress concentrator from which the crack may propagate, which may reduce the fatigue strength of the component. Sometimes, under harsh environmental conditions, such as in high temperatures as well as in heavy loading, surface defects may cause catastrophic failure of the component. Inferior fatigue properties, which are exhibited by irregular and rough surfaces, could be improved by the surface finishing. A good-quality surface free from any flaws, achieved from surface finishing, imparts higher fatigue strength and, in turn, higher fatigue life of the component.

b. *Corrosion resistance:* Due to providing more surface area, surfaces with higher surface roughness values are more vulnerable to corrosion [3]. Finished surfaces that are free from surface defects provide good resistance to corrosion as well as other chemical actions.
c. *Abrasion and wear resistance*: In many engineering applications, the mating surfaces have relative motions, due to which they are very prone to abrasion and wear. Frictional forces are higher in rough surfaces as compared to smooth surfaces under identical conditions. Due to less frictional forces, less heat is generated in smooth surfaces during relative motion, which leads to lesser wear in abrasion.
d. *Improve lubricity*: The lubricity or the capacity of a lubricant to reduce the friction is assessed by surface roughness and the viscosity of the oil. The lubricity of the oil is affected by the heat generated during the process. Since less frictional heat is generated in the relative motion of the smooth surfaces, the lubricity of the oil improves.
e. *Light reflectivity:* In many engineering applications such as in laser optics, high-reflective surfaces are of utmost importance. These light-reflective surfaces are generated either by coating the substrate materials or by finishing processes. The rougher surface scatters the light due to which the reflectivity of the rougher surface decreases, which again reduces the efficiency of the optical system due to these optical losses.
f. *Good aesthetic appearance:* A customer is always influenced by the aesthetic appearance of the product. If the product is aesthetically pleasing, it gives a positive impression on the customer and adds value to the product. A designer during designing a product gives considerable attention to its appearance, how it will look after manufacturing, along with other design parameters. For good aesthetic appearance, the product surface should be free from surface flaws such as blemishes, scratches, rigorous surface roughness, etc.
g. *Safety to users:* Safe operation and handling are a must from the user's point of view. Smooth surfaces give a sense of safety during the operation or handling. Smooth surfaces free from burrs, fine corners, scratches, and sharp edges confirm the safety of the users and offer pleasant feelings during handling.

1.3 FINISHING PROCESSES

A finishing process for a specific product is selected based on the application of that product. In some cases, a very high level of surface finish is required, such as in solid reflectors, but in other cases, less or marginal surface finish is required. Every finishing process is characterized by its maximum achievable finishing capacity, which this process can impart in its ultimate use. So, as per the application the finishing process is selected. Finishing is a post-processing operation and almost the last operation in some cases. If the finishing process is incorrectly selected, it could be a cause of rejection of the complete product. Contrary to this, selecting the appropriate finishing process can augment its characteristics with further value addition.

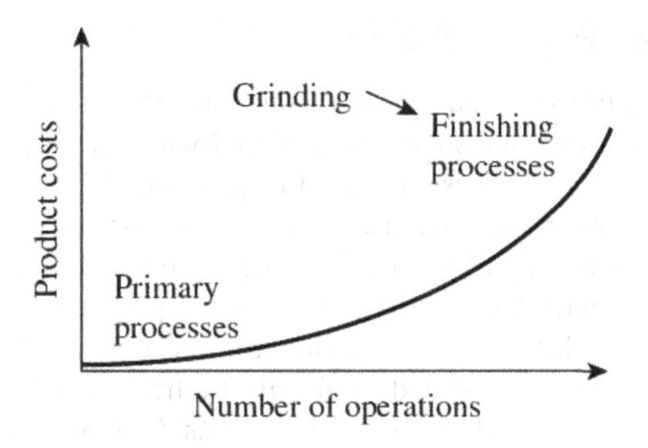

FIGURE 1.3 Product cost vs. number of processing operations.

Finishing demands specialized equipment, materials, and a high level of skill, which sometimes leads to higher cost of the operation. In some cases, it may alone account for 10–15% of total cost of production and in others multiplied by manyfold of production cost. Finishing operations require a high level of attention and extreme patience, the criticality of the finishing can be understood from the fact that even a single scratch can destroy the whole effort of finishing.

The cost of manufacture of a product is illustrated schematically in Figure 1.3.

Today's advanced (high-end) engineering applications need uniform finishing of complex and/or miniature components without altering the properties of surface or subsurface. This possesses a challenge to engineers and technocrats working in the field of finishing.

Abrasive-based surface-finishing techniques can be classified broadly into two categories: (i) traditional finishing processes and (ii) advanced finishing processes.

The detailed classification of abrasive-based finishing processes is shown in Figure 1.4 and discussed in the following subsections.

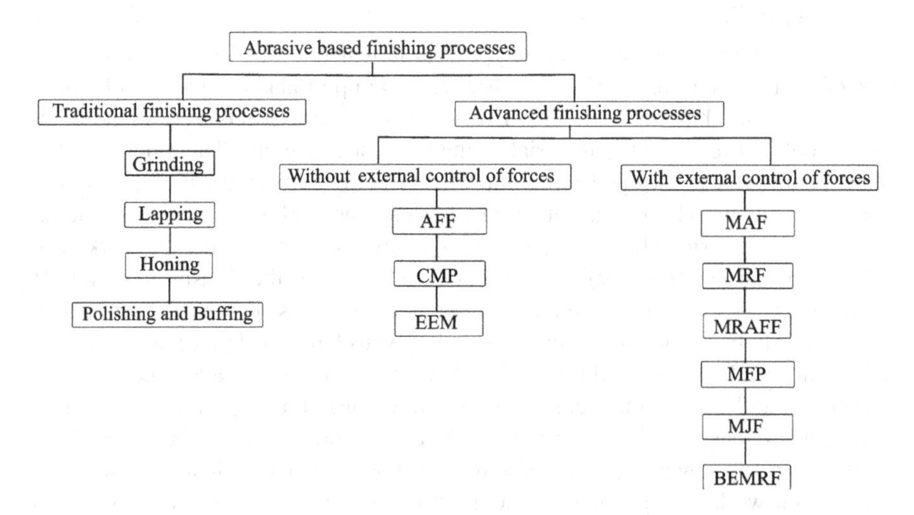

FIGURE 1.4 Classification of abrasive-based finishing processes.

1.3.1 Traditional Finishing Processes

Traditional finishing processes have been used in industry for a long time. Generally, these finishing processes use a solid finishing tool with simple tool geometry such as disc, cylinder, bar, etc. Processes named as grinding, lapping, honing, buffing, and superfinishing have been categorized in this category of traditional finishing processes. These processes are subjected to shape and size limitations by virtue of a solid (non-flexible) tool. Due to this reason, the lapping process is suitable for external flat surfaces, while honing is suitable for internal cylindrical surfaces. But complex surfaces cannot be finished uniformly by traditional finishing processes because of deployment of non-flexible finishing tools. If some special arrangements have been employed to finish the complex surfaces, it makes the process more complex and the extra cost of special arrangement is also incurred in finishing. Also, finishing complex surfaces by traditional finishing processes needs a high level of technical skill and work experience.

Finishing complex surfaces requires process automation, which is again difficult to achieve in traditional finishing processes. All these mentioned reasons make traditional finishing processes unsuitable for finishing complex surfaces.

In these processes, the finishing forces could not be controlled precisely, thus ultra-finished surfaces that require gentle finishing forces are difficult to achieve. Also, these processes generate a large amount of heat during their operation, which can be a cause of different types of surface defects due to excessive heating. Therefore, in traditional finishing processes extra care is required to precisely control the parameters responsible for heat generation. Grinding is one of the examples of finishing processes in which excessive heat is associated with the process that leads to severe surface defects. In these types of processes, continuous removal of heat from the finishing zone is must. Microcracks, residual stresses, warping, and thermal stresses are the major defects, which are produced due to excessive heat generation in the process.

1.3.1.1 Grinding

Grinding is a very common machining process in which a solid grinding tool in the form of a wheel is used. A grinding wheel is made up of abrasive particles binding together by some binding materials. Every abrasive particle acts as a single-point cutting tool and removes the material in the form of tiny chips. The randomly distributed abrasive particles present on the wheel's peripheral surface perform cutting action when the wheel is pressed and rotated against the workpiece. Since in grinding the cutting is performed by the relative motion between abrasives and the workpiece material, the hardness of the workpiece should be lower than the abrasive materials. If the hardness of the workpiece and the cutting material (abrasive) were the same, then there will only be rubbing between the grinding wheel and workpiece, and theoretically no material removal will take place. Since excessive heat is generated in grinding, coolant and lubrication become essential in the grinding and play an important role when a higher material removal rate (MRR) or good surface quality is required. In the past, many new grinding fluids and their application methods were developed.

Based on workpiece geometrical configurations, all the grinding processes can be classified into two broad categories: surface grinding and cylindrical grinding.

On the basis of grinding wheel-workpiece interactions, the grinding processes of four types are in practice as follows: (i) **Peripheral surface grinding:** in this type of grinding, the grinding wheel removes the material from the flat workpiece surface. In this configuration, the workpiece is mounted on the linear reciprocating table and is passed against the rotating grinding wheel, as shown in Figure 1.5(a). (ii) **Peripheral cylindrical grinding:** in this type of grinding process, the outside surface of the workpiece can be transformed in a variety of shapes, most commonly cylindrical, but others such as eccentric elliptical surfaces are also possible to generate. The schematic diagram of the process is shown in Figure 1.5(b). In peripheral cylindrical grinding, the workpiece must have an axis of rotation. During its operation, either the grinding wheel or the workpiece is given transverse feed. (iii) **Face-surface grinding:** face-surface grinding is a process in which the side face of the grinding wheel is used for the grinding purpose, Figure 1.5(c). In this process, the object to be ground is pressed against the side surface of the rotating wheel. This process is generally used for hard-to-reach surfaces and to sharpen the object. (iv) **Cylindrical-surface grinding:** cylindrical surface grinding is like face-surface grinding with the only difference that in face-surface grinding the workpiece is pressed against the side surface of the grinding wheel and there is no rotation of the workpiece, but in cylindrical-surface grinding the workpiece is also rotated, Figure 1.5(d).

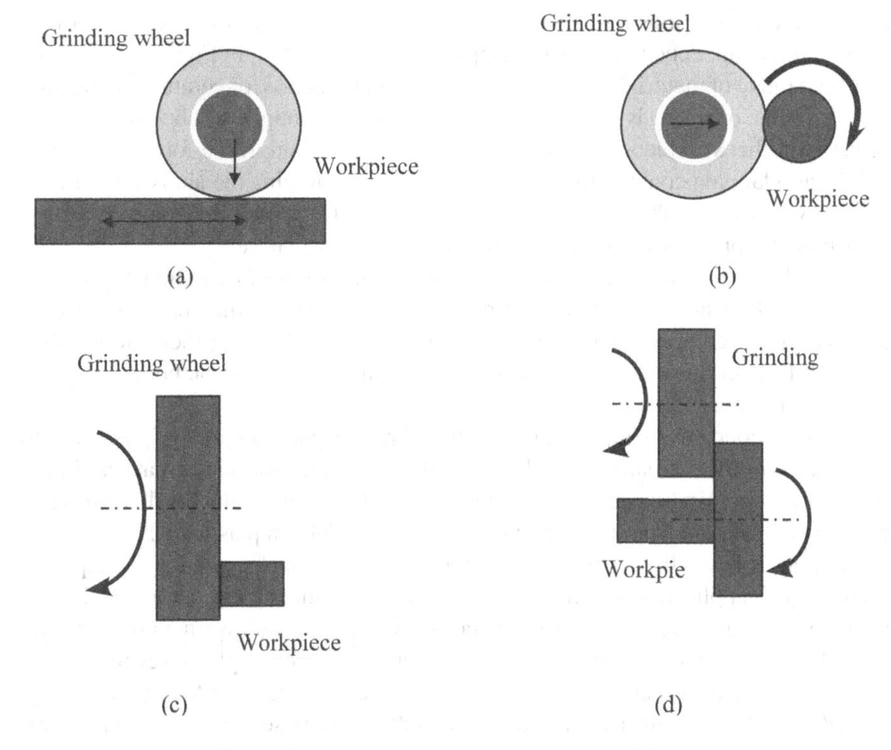

FIGURE 1.5 Schematic of basic grinding processes: (a) peripheral surface grinding, (b) peripheral cylindrical grinding, (c) face-surface grinding, and (d) face-cylindrical grinding.

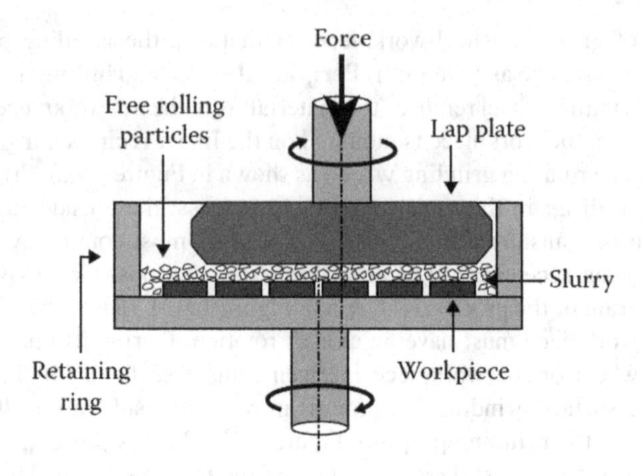

FIGURE 1.6 Lapping process with loose abrasive between lap and workpiece. (From ref. [4].)

1.3.1.2 Lapping

Lapping is a superfinishing process. It is fast compared to other finishing processes. It can impart a high degree of surface finishing and dimensional accuracy. Lapping is performed using loose abrasives. The abrasives are charged between the workpiece and the lap. Schematically the lapping process is shown in Figure 1.6.

Lap is made of materials like pitch, cast iron, copper, glass, or ceramic. Some fluid (oil, kerosene, or water) is used as the carrier medium to supply the abrasive particles in between the workpiece and the lap. Abrasive laden fluid in the form of slurry is named as a lapping compound. The fluid, along with carrying the abrasive particles, also provides some lubrication effect to make the cutting smooth. Commonly used abrasives in lapping compounds are alumina and silicon carbide.

In the lapping process, material is removed in the form of tiny chips by the two distinct mechanisms of abrasive cutting. In one mechanism, the abrasive particles are free to move between the workpiece and the lap, and during their motion they roll and slide. In this mechanism, they perform marginal cutting on both workpiece as well as on lap.

In the second mechanism, some of the abrasive particles are gripped due to embedding in the soft surface of the lap and remove the workpiece material in the same way as in grinding action. The mechanism of lapping on the brittle material is completely different from the lapping on metals that deform plastically.

Lapping is basically performed where a high level of surface quality is required. Some of the applications of this process are in manufacturing of optical lenses, measuring gauges, metallic bearing surfaces, metallographic specimen preparation, optical flats, and in other measuring instruments. This process improves the fatigue properties so that the components that are subjected to fatigue loading are often lapped. The lapped surfaces are good for sealing purposes also. Lapping operations can be performed on a wide variety of materials such as glass, steel, cast iron, aluminum, copper, etc. Lapping is performed manually, but for high accuracy and precision, lapping machines are also used. In machine lapping of soft materials, lapping pressure is kept in the range of 0.01–0.03 N/mm^2, and for hard materials the lapping pressure is kept as 0.07 N/mm^2 [5]. Lapping is primarily for geometric form correction; along with form correction, it also reduces the surface roughness.

1.3.1.3 Honing

Honing is primarily used to correct the geometric form of the surface, however, it also improves the surface finish. It uses bonded abrasive sticks mounted on a tool. Based on the circumferential surface area of the tool, the numbers of sticks mounted on the tool are decided.

This process is suitable for internal cylindrical surfaces like internal combustion engine bore, hydraulic cylinders, air-bearing spindles, gun barrels, etc. The sticks, composed of bonded abrasives, are pressed against the workpiece surface, and moved over it for the cutting action. The honing tool is given two motions, one axial motion along the length of the workpiece and a second rotary motion. During the motion of the tool this ensures that a particular point on the abrasive stick does not repeat its path multiple times. While finishing, the abrasive sticks should cover the entire length of the workpiece and should not leave the workpiece surface at any point of time.

Honing is considered the superfinishing process, and it gives surface finish in the range of 0.13–1.25 μm as compared to 0.9–50 μm in grinding. Honing is normally performed after precision machining, such as grinding, to get desired surface characteristics. Honing is an abrasive finishing process that is used for geometrical corrections, as well as for improving the dimensional accuracy of the functional surfaces of engineering components.

The flow of coolant is always maintained in the honing process to carry away the chips and the heat generated in the process. In some applications where proper lubrication is required, cross-hatched surfaces are generated by a honing process. This cross-hatched pattern retains the lubricating oil for the longer duration. The schematic of the honing process and the cross-hatched finishing pattern are shown in Figure 1.7.

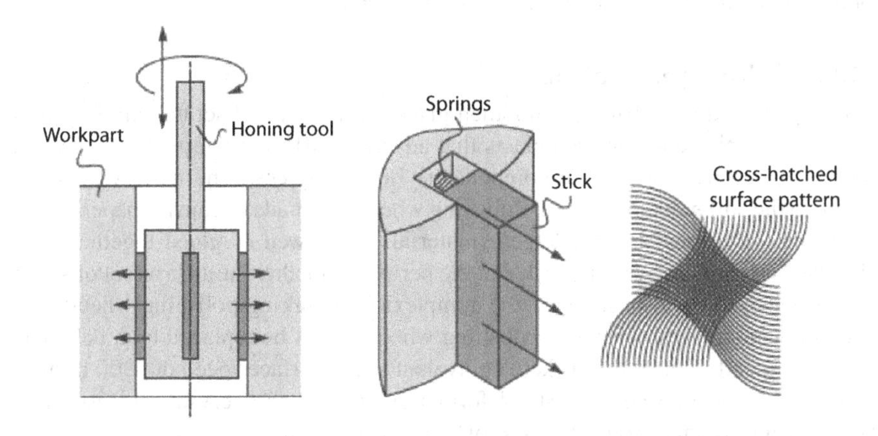

FIGURE 1.7 Schematic of honing process and cross-hatched surface pattern. (From ref. [4].)

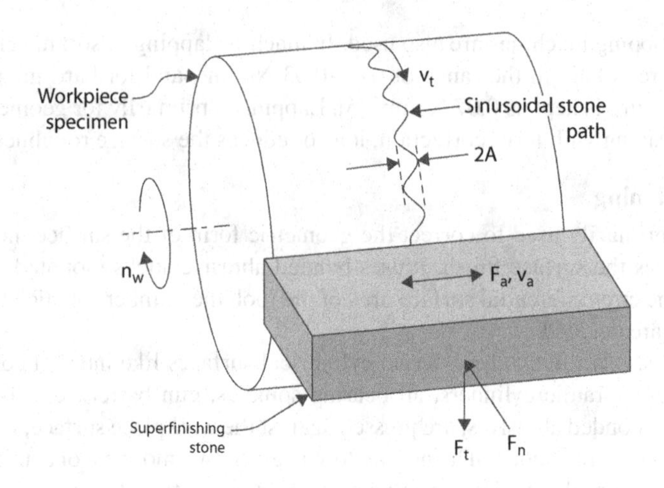

FIGURE 1.8 Schematic of superfinishing process. (From ref. [6].)

1.3.1.4 Superfinishing

As the name implies it is a superfinishing process; in its working it is somewhat like honing. Contrary to honing, in this process a single finishing stick made of bonded abrasives is used in place of multiple finishing sticks as in honing (Figure 1.8).

The solid abrasive stick is given reciprocating motion with high frequency and low amplitude. The abrasive grit size and the pressure applied on the sticks are kept smaller as compared to honing [6]. Thus, the superfinishing process is merely for improving the surface finish and not meant for dimensional correction. In this process, a mirror-like surface finish could be achieved with a roughness value around 0.01 μm. Coolant is fed during the finishing operations to wash away the tiny chips from the tool-workpiece interface and to cool the workpiece.

Some applications of the superfinishing process are finishing of memory drums of computers, brake drums, pistons and cylinders, parts of sewing machines, bearing components, pins, punch and dies, shafts, axles, etc.

1.3.1.5 Polishing and Buffing

Polishing is used to remove or smoothen grinding marks, tool scratch marks, mold parting lines, pits, and surface defects that adversely affect the appearance and the functional performance of the components. In this process, the abrasives firmly adhere to the soft polishing wheel. Polishing wheels are made of canvas, paper, cloth, and leather. Different layers of these materials are sewed or glued together. The adhesive and the abrasives are coated at the periphery in alternating layers. Polishing belts are also in trend because of the complexity in making polishing wheels, but there are some places where the polishing wheel cannot be replaced by a belt. The rotational speed of the wheel is kept equivalent to the surface speed of 2300 m/min. This process performs some plastic deformation of the surface crystals due to which the high sites are made to fill low sites while finishing. The polished surface usually has a finish of 0.4 μm.

The lapping and polishing differ in the sense that in polishing a negligible amount of material is removed, and due to plastic deformation, it produces a brighter shiny surface, whereas in lapping, material is removed from the surface due to which bright, shiny surfaces are not produced. The polishing process usually follows grinding and precedes buffing.

Buffing is almost like polishing, but in buffing the wheels are softer than the polishing wheel, and the abrasives are loosely attached to the wheel [7]. Buffing produces smooth and reflective surfaces. The workpiece is pressed against a buffing wheel, which is charged with a suitable buffing compound. The buffing tools are also called *buffing rouges*. Buffing is attributed to mirror-like surface finish. Its industrial applications are the finishing of molds and dies, shafts, busses, furniture, metallic handles, automobile components, household utensils, decorative artifacts, bicycle parts, etc.

1.3.2 Advanced Finishing Processes

In advanced engineering applications, the requirements of designers are very high. Due to the advent of new materials, there is a requirement to machine and finish them precisely. The demand for three-dimensional, miniature, and ultra-finished components are very high in today's advanced industries dealing in molds, dies, aerospace, defense, medical, electronics, and optics, etc. Traditional finishing processes have two major problems in their applications. First, they have shape and size limitations due to which they cannot finish complex 3D surfaces; second, traditional finishing processes cannot fulfill the requirement of nano- or ultra-finishing. The prime requirement of nano-finishing is to finish with the gentle finishing forces that can be achieved only through the precise control of the finishing process. The finishing forces in traditional finishing processes are less controllable, therefore traditional finishing processes cannot be used to achieve a high level of surface finish. Another point to consider is that in traditional finishing processes the high value of finishing forces can lead to surface and/or subsurface damage. Hence, there is a need to explore some other finishing processes that are deterministic in nature, in which there is a provision to precisely control the finishing forces by some external means even during the process that is called in-process control of finishing forces. Those finishing processes are capable of finishing complex-shaped components without much alteration in the process equipment. These can impart optical level of surface finish, which are not possible by traditional finishing processes [8]. Achieving a high level of surface finish is always associated with the cost of production; as the surface-finish level increases, the cost of production increases. It has been observed that achieving surface finish of the order of less than 1 μm increases the cost of production of surface finish sharply [9].

All these demands led the researchers to develop other new finishing processes keeping in mind to overcome the limitations of traditional finishing processes. These finishing processes are called *advanced finishing processes*. To be very specific, only advanced finishing processes based on abrasive cutting are being discussed here.

Based on the fact that either the finishing forces are being controlled or not, abrasive-based advanced finishing processes can be classified into two major categories: (a) finishing processes without control of finishing forces and (b) finishing processes with external control of finishing forces [10, 11]. These processes are discussed in the following subsections.

1.3.2.1 Abrasive-Based Advanced Nano-Finishing Processes without External Control of Finishing Forces

In this category of nano-finishing processes, the finishing forces on the workpiece cannot be controlled by external means. Processes such as abrasive flow finishing (AFF), chemical-mechanical polishing (CMP), and elastic emission machining (EEM) have been included in this category where the finishing forces cannot be controlled very precisely. Due to this, these processes are indeterministic in their working. These processes are briefly discussed in the following subsections.

1.3.2.1.1 Abrasive Flow Finishing

The working of abrasive flow finishing (AFF) process is based on an abrasive mixed flexible media. The pressurized abrasive-laden viscoelastic media is passed through the surface to reach the desired quality of surface finish. The process was developed in the 1960s in the United States to finish difficult-to-reach surfaces of critical components of hydraulic and fuel systems of aircraft to assist aerospace industries. AFF is used to polish, deburr, radius, and to remove recast layers.

The finishing media consists of a type of polymeric carrier and abrasive particles, such as alumina, silicon carbide, diamond, etc. [11]. In AFF, because of its flexibility, the finishing media acts as a "self-deformable stone," which adjusts its shape according to the surface profile of the workpiece. Extrusion dies, medical implants, aerospace components, and microfluidic channels in glass and ceramics can be finished using AFF. The process capability of AFF can be seen from the fact that, irrespective of surface complexities, it can improve surfaces up to 80–90%. Surfaces produced by any method such as metal cutting, casting, electric-discharge machining, etc., can be finished using AFF [12]. In this process, two opposite vertical cylinders are used, and the finishing media is extruded forward and backward through the narrow space between the workpiece and fixture. Extrusion pressure and the cross-sectional area of the restricted space between workpiece and fixtures are the major parameters on which the finishing intensity of the process depends. As the extrusion pressure and the cross-sectional area of the restricted space vary, the rheological properties of the finishing media vary, which in turn affects the cutting action of the abrasive particles. The schematic diagram of the process is shown in Figure 1.9.

This process can finish a broad range of engineering materials, from soft to very hard, ferrous to non-ferrous and nonmagnetic to magnetic. The performance of AFF process depends on a number of process parameters those can be classified in three categories: (i) AFF parameters, (ii) parameters related to polishing media, and (ii) workpiece-related parameters.

The first category, AFF parameters, consists of extrusion pressure, number of cycles, size of restricted passage, flow speed, and machining time. The second category of parameters related to finishing medium consists of viscosity of the medium, abrasive type, abrasive grit size, and its concentration. The third category, namely workpiece-related parameters, consists of workpiece shape, hardness, pre-machining processes, and workpiece orientation for surface texture [12].

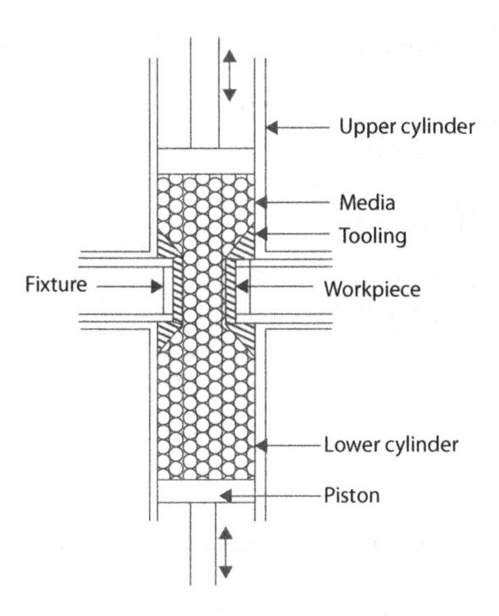

FIGURE 1.9 Abrasive flow finishing (AFF) process. (From ref. [12].)

1.3.2.1.2 Chemical-Mechanical Polishing (CMP)

Chemical-mechanical polishing (CMP) is a planarization process in which a combination of chemical reaction and abrasive cutting action is used to remove material from the workpiece surface. This process is widely used in manufacturing industries dealing in semiconductors. In this process, materials from the high points on the surface are removed gradually, and a flat surface is achieved. This process utilizes a chemical slurry-based polishing medium, which consists of carrier fluid, abrasive particles, and additives. During finishing, the workpiece is pressed against a porous, soft polishing pad generally made of polyurethane. The workpiece is held upside down, and the chemical slurry is supplied on the polishing pad, which meets the workpiece surface [13].

During finishing, first the workpiece surface reacts with the chemical present in the slurry and a passivation layer of the nanometer-level thickness is generated at the workpiece surface, which is either softer than the workpiece material (for hard workpiece materials) or harder than the workpiece material (for soft workpiece materials). Further, the abrasive particles remove this passivation layer by the abrasion. Formation and removal of the passivation layer continue throughout the finishing process, which gradually removes the workpiece material. In the CMP process, the major process parameters are slurry composition, abrasive material, abrasive particle size, polishing pad material, polishing pressure, workpiece, and polishing pad spinning speeds [14]. In semiconductor industries, CMP is used to polish and planarize silicon wafers, copper, and aluminum interconnects. In metal optics industries, it is used for ultra-finishing of metallic mirrors. A schematic diagram depicting the CMP process is shown in Figure 1.10.

A similar variant of this process is chemo mechanical polishing. In this process, the chemically active abrasives are used in polishing media in place of mixing chemicals separately. Abrasive particles chemically react with the workpiece surface, and

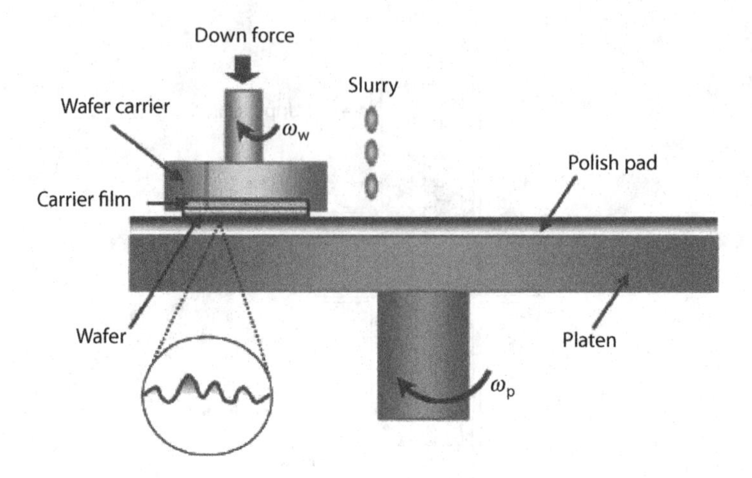

FIGURE 1.10 Schematic of chemical-mechanical polishing (CMP) process. (From ref. [16].)

the mechanical forces remove the reaction products [15, 16]. This process overcomes many problems associated with the CMP, such as surface damage and scratching due to hard abrasives, pitting on the surface due to brittle fracture, etc., resulting in smooth, defect-free surface.

1.3.2.1.3 Elastic Emission Machining (EEM)

In this process, the material is removed in the form of a single atom or cluster of atoms by the mechanical action. In this process, high-speed abrasive particles of very fine size strike with the atoms of the material to be finished either on individual atoms or a cluster of atoms and remove them from the parent materials. As the abrasive particles strike on the surface of the material, the material experiences an elastically-induced fracture without experiencing plastic deformation at the atomic scale [17].

Conceptually it could be assumed that when ultra-fine abrasive particles moving with the flow of any liquid strike with the surface atoms of the material to be removed, without pressing these particles over the workpiece surface, they remove the material in the form of clusters of atoms from the surface, as shown in Figure 1.11.

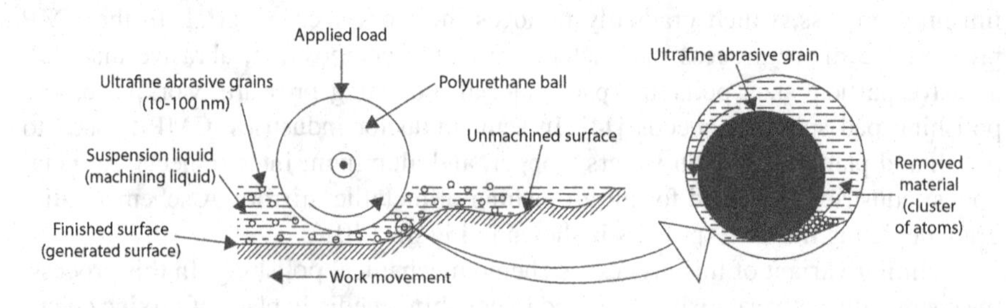

FIGURE 1.11 Schematic of elastic emission machining. (From ref. [11].)

The EEM produces a mirror-like surface free of crystallographic defects. In EEM, the surface finish achieved is approximate to the atomic size, i.e., 0.2–0.4 nm [17].

1.3.2.2 Abrasive-Based Advanced Nano-Finishing Processes with External Control of Finishing Forces

In this category of nano-finishing processes, the finishing forces applied on the abrasive particles can be controlled externally. Due to external control over the finishing forces, these processes are categorized as the deterministic finishing processes. These processes are basically magnetic field-assisted finishing processes in which the magnetic field is the main controlling factor. By controlling the magnetic field, the forces applied on the abrasive particles can be controlled, which, in turn, control the finishing action on the workpiece surface. In these processes, the finishing action of the abrasive particles can be varied even during the process by varying the applied magnetic field. These processes are capable of finishing up to the nanometer or even angstrom level by precisely controlling the finishing factors such as magnetic field, size of abrasive particles, size of magnetic particles, viscosity of the carrier medium, etc. Magnetic field-assisted finishing processes include magnetic abrasive finishing (MAF), magnetorheological finishing (MRF), magnetorheological abrasive flow finishing (MRAFF), and magnetic float polishing (MFP). These processes are discussed briefly in the following subsections.

1.3.2.2.1 Magnetic Abrasive Finishing (MAF)

Generally, rigorous penetration of the abrasive particle into the workpiece surface leaves deep scratches and indentations, which lead to surface damage. In the MAF process, it is possible to apply marginal finishing forces on the abrasives due to external control over the finishing forces. In MAF, the finishing forces are controlled magnetically by controlling the applied magnetic field [18, 19].

In the MAF process, either the abrasives are sintered with the ferromagnetic particles or sometimes the abrasive and ferromagnetic particles are mixed in loose/unbonded forms. When the abrasive particles are sintered with ferromagnetic particles, these sintered abrasives are termed as ferromagnetic abrasive particles. These sintered abrasive particles are also called magnetic abrasive particles and abbreviated as MAP [11].

When a magnetic field is applied, the MAP polarize magnetically; in this case, every particle behaves like an independent small permanent magnet. Under the magnetic field all these small magnetic particles cluster and bond together to shape into a flexible magnetic abrasive brush (FMAB).

An FMAB is composed of countless sharp-edged abrasive particles and acts like a multipoint cutting tool like the grinding tool. Each abrasive particle is equipped with multiple cutting edges. When the flexible abrasive brush is moved over the workpiece surface, the multiple cutting edges of the abrasive particles perform cutting actions and remove the workpiece material in the form of micro-sized chips from the workpiece surface [19, 20]. Slowly and gradually the unevenness of the workpiece surface is removed.

The magnetic field works like a binder and holds magnetic abrasive particles firmly in the working gap. Usually, the working gap is kept between 1–3 mm. When

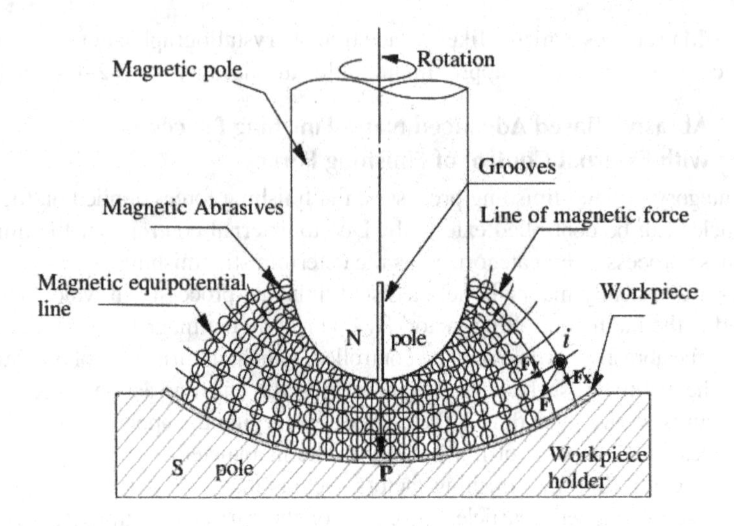

FIGURE 1.12 Schematic representation of MAF process. (From ref. [21].

this flexible brush is pressed on the workpiece surface, it applies a marginal normal finishing force by virtue of which the abrasive particles penetrate the workpiece surface. The relative motion of the brush provides the necessary shearing forces to the abrasives due to which the penetrated abrasives remove the workpiece material in the form of small chips on the micro/nano level.

Generally, the electromagnet is suggested to generate real-time change in magnetic field, but due to complexities involved in the manufacturing of electromagnets and their rotation, permanent magnets are also used. Where a permanent magnet is used to provide the magnetic field, there are two different methods of controlling the magnetic field. Either the magnetic field can be controlled by replacing the permanent magnet or by varying the working gap [19].

MAF can be used equally for finishing of external, internal, simple, or complex surfaces [21]. The schematic representation of MAF is shown in Figure 1.12.

MAF is a highly efficient process, and the performance of the process depends on intensity of the magnetic field, gap space between the tool and workpiece, rotational speed of workpiece or MAF tool, properties of workpiece material, content of ferromagnetic particles, and abrasive-related parameters such as type of abrasive, size of abrasive grit, and volume fraction in the mixture [21].

1.3.2.2.2 Magnetorheological Finishing (MRF)

Magnetorheological finishing (MRF) is a nano-finishing process in which the magnetic forces can be controlled by the application of magnetic fields. This process was developed by the Center for Optics Manufacturing (COM) situated in Rochester, New York, in 1996 with the invention and construction of a vertical wheel-based machine [22]. This process was developed to automate the finishing of high-precision lenses, those generally made up of brittle materials and very prone to cracks while finishing by traditional finishing techniques. The process was commercialized by QED Technology, Inc. in 1998 with the MRF machine designated as Q22 [9, 22].

The key element of the magnetorheological finishing process is magnetorheological polishing (MRP) fluid. MRP fluid is a smart material composed of fine abrasive particles, magnetizable particles of micron size, some carrier medium mineral oil, and silicon oil or water. Carbonyl iron particles (CIPs), due to their higher magnetic permeability and spherical shapes, are most used as the magnetizable particles in magnetorheological finishing. Carrier medium is a nonmagnetic viscous fluid. Some additives are also used to prevent the sedimentation of the particles in the fluid and to impart some other properties, such as corrosion resistance, etc.

In an unenergized state (when a magnetic field is not applied), the MRP fluid seems like paint and exhibits Newtonian behavior. In this state, the abrasive particles and CIPs are randomly scattered in the fluid volume. But, when the MRP fluid is subjected to a magnetic field environment, the CIPs get polarized, and north and south poles develop in the individual particles. The polarized particles get aligned by touching the opposite poles of the particles along the magnetic line of force. The attraction force between opposite poles of the particles imparts stiffness to the chains of the CIPs, and finally it reflects in the form of high viscosity of the fluid. The energized MRP fluid acts as the Bingham plastic fluid. The abrasive particles are entangled and gripped firmly in between CIPs' chains. When the stiff MRP fluid interacts with the workpiece surface and they both are subjected to the relative motion, the normal force acts on the abrasive particles through the CIPs and results in the penetration of abrasive particles in the workpiece surface. The shearing force caused by the relative motion between the workpiece and the stiff MRP fluid impart the shearing action due to which the penetrated abrasive particles remove the material from the workpiece surface in the form of small chips of micron size.

QED technology developed a wheel type of MR finishing setup [9]. In this setup, MR polishing fluid is deposited over a rim of a rotating wheel by the nozzle at one side of the wheel's periphery, and at the other side the MR polishing fluid is removed by suction (Figure 1.13).

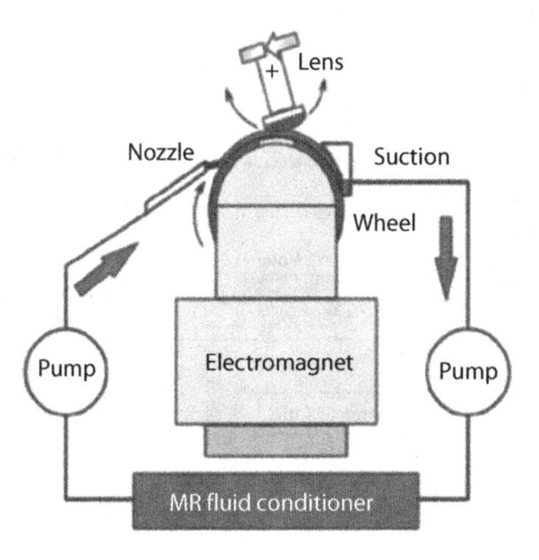

FIGURE 1.13 Schematic diagram of MRF setup. (From ref. [9].)

The wheel-shaped tool is magnetized due to which the MR polishing fluid takes the form of a stiffened ribbon. At the time of the wheel's rotation, the stiffened ribbon of MR polishing fluid is dragged between a gap formed by wheel periphery and the workpiece surface. The gap is converging in its cross-section because of the circular profile of the rotating wheel. Due to the narrow gap, the workpiece is subjected to a normal force exerted by the stiffened MRP fluid. The shearing force to the abrasive is provided by the wheel rotation due to which the penetrating abrasives removed the workpiece material in a gradual manner. This process is generally used for finishing of brittle materials such as ceramics, optical glass, etc., but also for finishing of plastics and nonmagnetic metals [11].

MRF has the capability to finish optical lenses in the range of 10–100 nm on the scale of peak to valley height, and on a root mean square (RMS) scale it is 0.8 nm by overcoming many difficulties associated with traditional finishing processes [22].

1.3.2.2.3 Magnetorheological Abrasive Flow Finishing (MRAFF)

Magnetorheological abrasive flow finishing (MRAFF) process is a combination of two different finishing processes, which leads this process to the category of hybrid finishing processes. MRAFF is the hybridization of magnetorheological finishing and abrasive flow finishing in which determinism of MRF and versatility and adaptability of AFF have been incorporated in a single process [23].

In AFF, the fine abrasive mixed polymeric medium acts as the *self-deformable finishing stone* that can adjust its shape while finishing according to the shape of the workpiece by virtue of its flexible nature [23]. This process is capable of finishing intricate-shaped components (external or internal features) by overthrowing inherent shape limitations of almost all the traditional finishing processes. The schematic diagram of the MRAFF setup and the finishing mechanism are shown in Figure 1.14.

The abrading action in AFF depends on the extrusion pressure applied on the finishing medium and the rheological properties of the polymeric substance. The

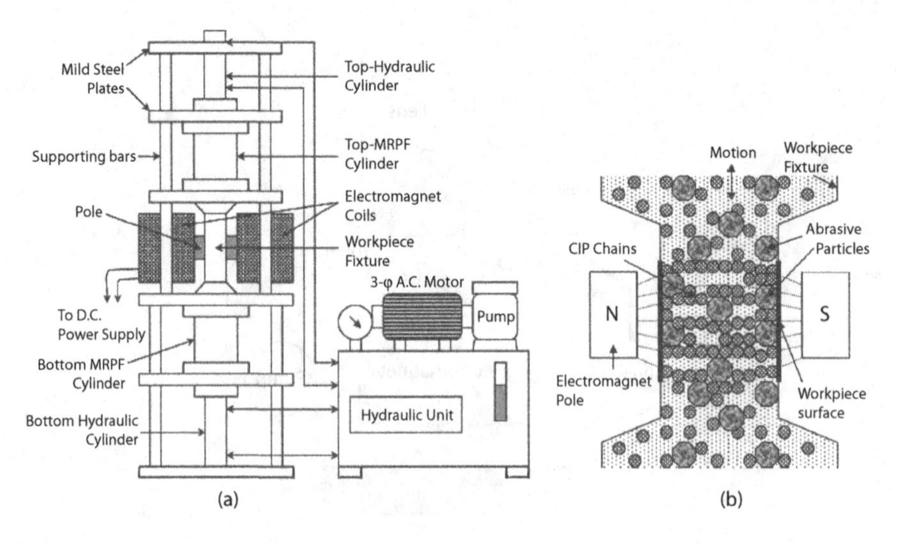

FIGURE 1.14 Schematic of (a) MRAFF setup, (b) mechanism of MRAFF. (From ref. [23].)

extrusion pressure can be controlled in AFF, but the rheological properties of the polymeric medium cannot be controlled by the external means. To introduce the determinism and to enhance the finishing effect (for hard-to-finish materials), MRAFF has been developed by utilizing the MRP fluid in place of nonmagnetic polymeric medium.

In MRAFF, the rheological properties of the finishing medium can be varied externally by varying the magnetizing current supplied to the electromagnet. Due to normal force on the abrasive particles through CIPs, abrasive particles penetrate the surface to be finished, and the extrusion pressure provides the necessary shearing action to the abrasive particles.

In the MRAFF setup, two opposite vertical hydraulic cylinders are used to retain the MRP fluid. The MRP fluid is pushed forward and backward from one hydraulic cylinder to another through the passage formed by the fixture using hydraulic actuators [23]. The workpiece is placed in between the hydraulic cylinders, with proper fixtures and tooling, in such a way that the pressured MRP fluid passes through or over the workpiece surface during the fluid's forward and backward motion. The MRP fluid is exposed to the magnetic field by utilizing opposite electromagnetic poles around the workpiece as shown in Figure 1.14(b).

1.3.2.2.4 Magnetic Float Polishing (MFP)

Among the traditional as well as advanced finishing processes, the majority of the finishing processes are for finishing of flat surfaces, cylindrical surfaces, or the combination of both leading to complex three-dimensional surfaces. Finishing of spherical surfaces are still challenging and less covered by the well-known finishing processes, but it is equally important in advanced engineering applications. Magnetic float polishing (MFP) is one such finishing process that has the capability to finish spherical surfaces. MFP is a magnetic field–assisted finishing process in which the finishing forces can be controlled as per the finishing requirement, which makes this process deterministic in nature. Spherical metallic or ceramic balls can be finished precisely by this process [24].

Magnetic fluid's ferro-hydrodynamic behavior is the basis on which MFP works. According to this behavior, on the application of magnetic fields the nonmagnetic substances suspended in magnetic fluid levitate. In MFP, the abrasive particles and nonmagnetic polishing float are the suspended nonmagnetic substances in the magnetic fluid [25]. As the magnetic field is applied, the region of the higher magnetic field attracts the magnetic particles due to which the magnetic particles start moving towards the higher magnetic field region. The movement of the magnetic particles results in a buoyancy force, which acts on the nonmagnetic substances (nonmagnetic float and abrasives) and forces them to move to the region of the lower magnetic field. This process is considered highly efficient because the levitation force is applied on the abrasive particles in a controlled way. Extremely small finishing forces, about 1N or less per ball, are applied by the abrasives [25].

In MFP, the magnetic field is applied by the north and south poles of the strong electromagnets arranged alternatively below the finishing container, as shown in Figure 1.15. In this arrangement, the finishing chamber is filled with a magnetic fluid containing polishing abrasives. In this abrasive-mixed magnetic fluid, balls to be finished are suspended in a batch of 10–100 [25].

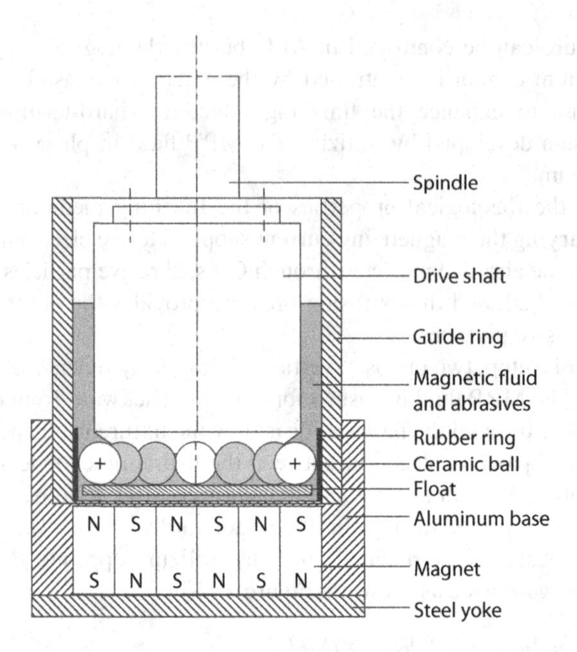

FIGURE 1.15 Schematic diagram of the magnetic float polishing apparatus [26].

In its working, as the magnetic field is imposed, all the elements in the magnetic fluid made of nonmagnetic materials such as abrasives, float, and balls to be finished start floating. The buoyancy force, due to the downward motion of the magnetic particles, pushes them upward in the magnetic fluid. The required force level is achieved by pressing the balls using a drive shaft. The drive shaft is rotated, which also rotates the balls and forces them into the magnetic fluid against the nonmagnetic float. The container is made up of a nonmagnetic material so that the magnetic force lines do not concentrate in the container but stiffen the magnetic fluid stored in it. The polishing is performed by the abrasives gripped in the magnetic particles of the stiffened magnetic fluid when the balls get rotated against the abrasive particles. By this process, it is possible to get superfinished surfaces free from surface flaws such as scratches, surface cracks, pits, etc., on spherical shaped components.

1.3.2.2.5 Magnetorheological Jet Finishing (MRJF)

Magnetorheological jet finishing (MRJF) technology is a variant of MR finishing processes, which is based on magnetorheological fluid. In MRJF, a jet of magnetorheological polishing fluid is used for the finishing purpose. This process is used to finish components with shape limitations due to which mechanical interference occurs during finishing. This process is suitable for finishing conformal and difficult-to-reach surfaces, such as steep slopes of concave optics, which are difficult to finish by traditional finishing processes. In this process, a jet of magnetorheological fluid containing abrasive particles is stabilized by an axial magnetic field, which results in a highly collimated, coherent jet ejected from the nozzle by suppressing the most dangerous initial disturbances [27–29].

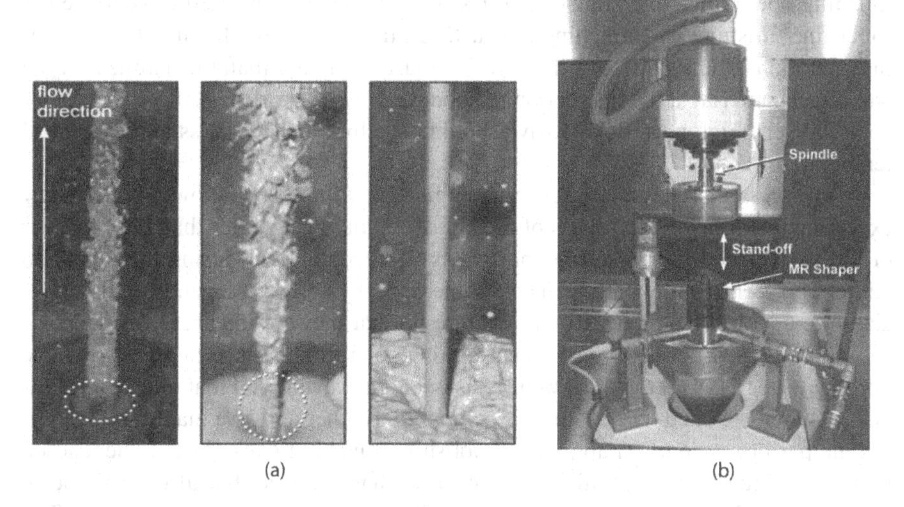

FIGURE 1.16 (a) Jet snapshot images (velocity = 30 m/s, nozzle diameter = 2 mm), (b) MRJF setup. (From ref. [28].)

When the round-shaped magnetically stabilized jet impinges on the workpiece surface, the roughness peaks present on the surface to be finished are sheared off by the abrasives due to the energy gain by the radial splash of the jet [28, 29]. Beyond the magnetic field, the stabilized structure of the jet induced due to the magnetic field starts decreasing and the jet flares out.

In the case of water, the jet remains stabilized only for almost two nozzle diameters. The MR fluid has a higher viscosity than the water due to which the MR fluid becomes more stabilized, and the coherent portion of the MR jet extends to almost 7–8 diameters. After the coherent portion of the jet, it breaks down and spreads in the form of droplets. As an axial magnetic field is applied, the jet of MR fluid remains coherent for more than 200 diameters [27–29]. The same is illustrated in Figure 1.16(a), and the setup is shown in Figure 1.16(b).

1.3.2.2.6 Ball-End Magnetorheological Finishing (BEMRF)

Ball-end magnetorheological finishing (BEMRF) is a magnetic field-assisted finishing process. This process is also a variant of MR finishing only with the difference that it has an axially rotated spindle made of ferromagnetic materials. This process was developed to finish complex surfaces, such as three-dimensional surfaces, intricate cavities, and difficult-to-reach surfaces. In detail, BEMRF is discussed separately in Chapter 3.

1.4 TYPES OF ABRASIVES

To impart the mirror-like surface finish on the product, the abrasive particles can perform the gradual cutting in small amounts by means of plastic deformation in place of brittle fractures, which generate rougher surfaces. In such polishing/

finishing operations, fine abrasives of less than 1-μm size are used to generate the super-fine finishing by scratching the surface microscopically. In any of the finishing operations, the hardness of the abrasive must be greater than the hardness of the workpiece material at the interaction temperature. In this regard, as this temperature could be very high now, the abrasive should have high hot hardness at the elevated temperature.

In all abrasive processes, the criterion of hardness should be followed without any exception, meaning the hardness of the abrasives must be greater than the hardness of the workpiece; otherwise the abrasives will become blunt during the operation without performing any cutting. The hardness of any abrasive depends on the interaction temperature; at elevated temperature the hardness of the abrasives decrease substantially. Most of the abrasives suffer almost half the harness at 1000°C. Cubic boron nitride (CBN) is the most wear-resistant abrasive material that retains its hardness better at elevated temperature as compared to other abrasive materials.

The performance of an abrasive in polishing/finishing does not only depend on hardness of the abrasive, which is a material characteristic, but also on shape of abrasive particles and their particle-size distribution in whole mass of abrasives. The shape of the abrasive particles depends on the material properties such as fracture toughness, cleavability, etc., and on the manufacturing process or processes utilized for producing them as well. Along with the shape of abrasive particles, the distribution of abrasive particles of different sizes that is known as grain-size distribution also plays a key role in polishing/finishing, and it is also a parameter to influence the outcome of the finishing.

The MRR, and the surface finish achieved are the function of abrasive size. As the abrasive size increases the MRR increases, but the surface finish decreases. The reverse is also true: as the abrasive size decreases the MRR decreases, and the surface finish increases. Lower MRR imparts the better surface finish. In surface finishing the main objective is to remove the surface roughness peaks gradually only, not to remove the material in bulk.

Abrasives are crystalline in nature. The abrasive can be categorized in two groups: natural abrasives and artificial abrasives. Natural abrasives are found in the earth's crust, and they include diamond, quartz, emery, corundum, garnet, etc. Artificial abrasives are manufactured abrasives such as silicon carbide (SiC), aluminum oxide (Al_2O_3), synthetic diamond, cubic boron nitride (CBN), boron carbide etc. The artificial abrasives, based on their production process can again be categorized into two groups: fused abrasives and unfused abrasives. Fused abrasives are produced at a very high temperature in the electric furnace, which results in hard crystals. Unfused abrasives are produced at lower temperature using some chemical additives due to which they do not exhibit hard crystalline structure as found in fused abrasives.

Five abrasive materials are most common in industries and widely-used in different types of grinding and polishing work; out of those five, three abrasive materials, namely silicon carbide, aluminum oxide, and garnet come under the category of conventional abrasive materials. The remaining two abrasive materials are diamond and cubic boron nitride; these are considered super abrasives based on their high hardness [30, 31].

The various types of abrasives and their specific applications are discussed as follows:

i. ***Diamond:*** Diamond is the hardest material available in nature. It has multiple randomly oriented sharp edges, which provide a matte surface finish. Both natural and synthetic abrasives are used in industries. It has a hardness of the order of 10 on the Mohs scale and is considered the hardest abrasive material. Considering the cost of the abrasive, diamond is used in some specific applications. Generally, it is used for machining/finishing of very hard materials, such as ceramic. It is generally not recommended for soft materials because of the problem of embedding of abrasive particles in the surface, but when super-fine finishing is required then very fine diamond abrasives can be used in a paste form. The rate of machining is faster with the diamond abrasives due to their fast-cutting action. Considering the cost and the availability of abrasives, synthetic diamond abrasives have been widely used in industrial applications over natural diamond abrasives [30, 31].
ii. ***Cubic boron nitride (CBN):*** This is an allotropic form of cubic nitride that is crystalline in nature. Cubic boron nitride (CBN) basically matches with the diamond in its hardness on the Mohs scale but is more thermal resistant than the diamond. Due to its thermal-resistant property, it can be used in application of higher temperature. Generally, it can work well at 1900°C (3500°F). Cubic boron nitride (CBN) possesses good chemical resistance to ferrous materials and due to which it does not carbonize in interacting with ferrous materials contrary to the diamond abrasives. High-speed machining as well as high stock removal is also possible with CBN in grinding. It is most suitable for super alloy, ceramics, and high grades of steels, such as tool steel, die steel, etc.
iii. ***Silicon carbide:*** This is a hard crystalline ceramic material. The abrasives of silicon carbide have sharp crystal edges. Silicon carbide has a hardness of 9.5 on the Mohs scale just after the hardness of diamond, cubic boron nitride, and boron carbide. It imparts a gray matte surface finish during the operation. During operation, the grains of silicon carbide break down and perform smoother and faster cutting. Silicon carbide abrasives are most suited for hard, friable, and low tensile-strength materials.
iv. ***Aluminum oxide (Al_2O_3):*** This is a cheaper and widely used abrasive material. Aluminum oxide is less friable or tougher than the silicon carbide. During their operation, they do not break easily at the applied load, so no fresh cutting edges are generated, which leads to the dullness of the abrasive particles. They possess hardness 9 on the Mohs scale, which is just after the hardness of silicon carbide. They are most suited for the grinding finishing of high tensile-strength materials and for rough work.
v. ***Garnet:*** This is a natural mineral mined from the earth. The property of friability of garnet is over the harness of these abrasives. Due to friability, these abrasives break rather than embedding in the surface at high loads. These abrasives are most suitable for applications where embedding of

abrasives is not desirable, such as grinding/finishing of gears, lapping of metallic seals, etc. In loose form these abrasives are used for mirror finishing on glass, and edge rounding of lenses, etc. They possess a hardness of 8–9 on the Mohs scale, putting them next to aluminum oxide.

REFERENCES

1. Palanikumar, K. (2012). Analyzing surface quality in machined composites. In H. Hocheng (ed.), Machining Technology for Composite Materials (pp. 154–182). Woodhead Publishing.
2. Lu, C. (2008). Study on prediction of surface quality in machining process. Journal of Materials Processing Technology, 205(1–3), 439–450.
3. Budke, E., Krempel-Hesse, J., Maidhof, H., & Schüssler, H. (1999). Decorative hard coatings with improved corrosion resistance. Surface and Coatings Technology, 112(1–3), 108–113.
4. Jain, V. K. (Ed.). (2016). Nanofinishing Science and Technology: Basic and Advanced Finishing and Polishing Processes. CRC Press.
5. Rao, P. N. (2013). Manufacturing Technology (Vol. 1). Tata McGraw-Hill Education.
6. Nakayama, K., & Hashimoto, H. (1995). Experimental investigation of the superfinishing process. Wear, 185(1–2), 173–182.
7. Jain, V. K. (2009). Advanced Machining Processes. Allied Publishers.
8. Sankar, M. R., Jain, V. K., & Ramkumar, J. (2009). Experimental investigations into rotating workpiece abrasive flow finishing. Wear, 267(1–4), 43–51.
9. Jain, V. K. (2009). Magnetic field assisted abrasive based micro-/nano-finishing. Journal of Materials Processing Technology, 209(20), 6022–6038.
10. Gov, K., Eyercioglu, O., & Cakir, M. V. (2013). Hardness effects on abrasive flow machining. Strojniski Vestnik/Journal of Mechanical Engineering, 59(10).
11. Jain, V. K. (2008). Abrasive-based nano-finishing techniques: an overview. Machining Science and Technology, 12(3), 257–294.
12. Jain, R. K., & Jain, V. K. (2000). Optimum selection of machining conditions in abrasive flow machining using neural network. Journal of Materials Processing Technology, 108(1), 62–67.
13. Zhong, Z. W., Wang, Z. F., & Zirajutheen, B. M. P. (2005). Chemical mechanical polishing of polycarbonate and poly methyl methacrylate substrates. Microelectronic Engineering, 81(1), 117–124.
14. Chen, C. C. A., Shu, L. S., & Lee, S. R. (2003). Mechano-chemical polishing of silicon wafers. Journal of Materials Processing Technology, 140(1–3), 373–378.
15. Komanduri, R., Lucca, D. A., & Tani, Y. (1997). Technological advances in fine abrasive processes. CIRP Annals, 46(2), 545–596.
16. Kim, T. W., & Cho, Y. J. (2006). Average flow model with elastic deformation for CMP. Tribology International, 39(11), 1388–1394.
17. Mori, Y., Yamauchi, K., & Endo, K. (1987). Elastic emission machining. Precision Engineering, 9(3), 123–128.
18. Singh, D. K., Jain, V. K., & Raghuram, V. (2004). Parametric study of magnetic abrasive finishing process. Journal of Materials Processing Technology, 149(1–3), 22–29.
19. Jain, V. K., Kumar, P., Behera, P. K., & Jayswal, S. C. (2001). Effect of working gap and circumferential speed on the performance of magnetic abrasive finishing process. Wear, 250(1–12), 384–390.
20. Wang, Y., & Hu, D. (2005). Study on the inner surface finishing of tubing by magnetic abrasive finishing. International Journal of Machine Tools and Manufacture, 45(1), 43–49.

21. Lin, C. T., Yang, L. D., & Chow, H. M. (2007). Study of magnetic abrasive finishing in free-form surface operations using the Taguchi method. The International Journal of advanced Manufacturing Technology, 34(1), 122–130.
22. Kordonski, W. I., & Jacobs, S. D. (1996). Magnetorheological finishing. International Journal of Modern Physics B, 10(23n24), 2837–2848.
23. Jha, S., & Jain, V. K. (2004). Design and development of the magnetorheological abrasive flow finishing (MRAFF) process. International Journal of Machine Tools and Manufacture, 44(10), 1019–1029.
24. Umehara, N., Kirtane, T., Gerlick, R., Jain, V. K., & Komanduri, R. (2006). A new apparatus for finishing large size/large batch silicon nitride (Si3N4) balls for hybrid bearing applications by magnetic float polishing (MFP). International Journal of Machine Tools and Manufacture, 46(2), 151–169.
25. Raghunandan, M., Umehara, N., Noori-Khajavi, A., & Komanduri, R. (1997). Magnetic float polishing of ceramics. Trans. ASME; Journal of Manufacturing Science and Engineering, 119, 520–528.
26. Komanduri, R., Hou, Z. B., Umehara, N., Raghunandan, M., Jiong, M., Bhagavatula, S. R., ... & Wood, N. O. (1999). A "gentle" method for finishing Si 3 N 4 balls for hybrid bearing applications. Tribology Letters, 7(1), 39–49.
27. Kordonski, W. I., Shorey, A. B., & Tricard, M. (2006). Magnetorheological jet (MR Jet TM) finishing technology. Journal of Fluids Engineering, 128(1), 20–26.
28. Tricard, M., Kordonski, W. I., Shorey, A. B., & Evans, C. (2006). Magnetorheological jet finishing of conformal, freeform and steep concave optics. CIRP Annals, 55(1), 309–312.
29. Tracy, J., Kordonski, W., Shorey, A., & Tricard, M. (2007, May). Advances in finishing using magnetorheological (MR) jet technology. In Optifab 2007: Technical Digest (Vol. 10316, pp. 103160F). International Society for Optics and Photonics.
30. Marinescu, I. D., Uhlmann, E., & Doi, T. (Eds.). (2006). Handbook of Lapping and Polishing. CRC Press.
31. Marinescu, I. D., Rowe, W. B., Dimitrov, B., & Ohmori, H. (2012). Tribology of Abrasive Machining Processes. William Andrew.

2 Magnetic Abrasive Finishing

2.1 INTRODUCTION

Magnetic abrasive finishing (MAF), which is a high-precision super-finishing process, initially was developed in United States in the early 1940s. But further development was carried out in former U.S.S.R. and Bulgaria in the late 1950s and 1960s. Soviet scholars, such as Baron and his associates [1, 2–5], and Bulgarian scholars, Makedonski and his team of researchers [3], applied this technique to a variety of products and applications. Later on, in the 1980s, further research on finishing and polishing, using the same technique, was carried out by Japanese scholars at different universities of Japan such as at Tohoku University by Kato and his team of researchers [6], at Tokyo University by Nakagawa and his research team [7], and at Utsunomiya University by Shinmura and his research team [8–11]. Shinmura et al. [9] continued their research work in the same domain and developed various equipment for finishing/polishing of products of different configurations, such as for finishing of the internal surface of a tube, external surfaces of cylindrical components, and finishing of flat prismatic surfaces. Dr. Shinmura, in all his work, performed the studies on ferromagnetic materials.

Since its development, MAF has been used for finishing of high-precision and sensitive instruments or products used in medical, electronics, optics, die and mold manufacturing, aerospace, and engine components [12].

In MAF, either the magnetic abrasives or the loose abrasive particles mixed with some types of ferromagnetic particles are used. Magnetic abrasives are magnetizable particles of small size with grinding ability. These types of particles magnetize when exposed to a magnetic field; the particles cluster due to the magnetic line of forces and form a flexible brush. When a relative motion is provided to this flexible brush, the brush acts like a flexible grinding tool and removes undulations from the surface to be finished. The magnetic abrasive particles or a mixture of loose abrasives and magnetizable particles are filled in the working gap formed between the workpiece surface and the magnetic poles. For the required magnetic field, either permanent magnetic poles are used, or an electromagnet is used to create magnetic poles. In the latter case, the intensity of the magnetic field can be controlled as per the finishing requirement, but in case of poles made by permanent magnet, the intensity of the magnetic field and hence the finishing rigorousness cannot be controlled.

The flexible brush of magnetic abrasive particles or the mixture of abrasive particles and magnetizable particles adjust its compliance according to the shape of the workpiece to be finished/polished. The flexible brush, having some stiffness due to the influence of the magnetic field, exerts finishing forces at the workpiece surface.

DOI: 10.1201/9781003228776-2

Under the action of finishing forces, if the relative motion is provided to the workpiece and the flexible brush, the phenomenon of microgrinding occurs, due to which the surface material is removed in the form of microscopic chips. Sometimes, the carrier fluid mixed with the magnetic abrasives is electrolytic in nature and this is done deliberately to promote the electrochemical reactions in the working gap. In such cases, along with abrasive wear, electrochemical wear also happens and contributes to the aggregate material removal [13]. The gradual removal of the material in the form of tiny chips improves the geometrical characteristics such as surface roughness, dimensional tolerance, etc. It also improves the physical and mechanical characteristics of the surface, such as load-bearing capacity of the surface, and enhances the life of the component by removing the points of stress concentrations, such as scratch and indents, etc. The overall quality of the surface is improved by enhancing the integrity of the surface. High-quality surface finish imparts good aesthetic appearance as well as good service performance [14].

Since the finishing forces in the MAF process are marginal, the chances of occurrence of defects, such as microcracks during the operation, are very low, especially in the case of brittle materials, which are very prone to crack formation. Primarily, cylindrical and flat surfaces are most suitable surfaces for finishing by MAF process. The material removal rate (MRR) of 1 μm/min and least surface roughness (Ra) of 5 nm can be achieved by this process [15].

By changing the finishing tool configuration, magnetic flux density, type of magnetic field generator, and relative motion between tool and workpiece in the MAF process, different types of surfaces can be finished, such as external cylindrical surfaces, internal cylindrical surface, flat surfaces, freeform surfaces, etc. The different finishing tool configurations, poles settings, and finishing methods are shown in Figure 2.1.

The classification of MAF can be done based on types of magnetic field generation, magnetic nature of the workpiece material, workpiece configuration, condition of finishing media, and nature of finishing media. This is depicted in Figure 2.2.

2.2 ESSENTIAL ELEMENTS OF MAF

Various essential elements of MAF are discussed in subsequent text.

2.2.1 Magnetic Field Generators

Magnetic field generators are nothing but permanent magnets or electromagnets. The magnetic lines of forces align the magnetic particles and work as the adhesive and impart some strength to the finishing lap. Due to the magnetic forces in the finishing zone, the magnetic abrasives apply finishing forces on the workpiece surface. In case of a permanent magnet, the finishing forces cannot be controlled because of the fixed magnetic flux density in the working gap, but in case of an electromagnet, the finishing forces can be controlled by varying the different parameters that control the generation of magnetic flux in the working gap such as magnetizing current, size of the electromagnet, and number of turns of the current wire on the electromagnet.

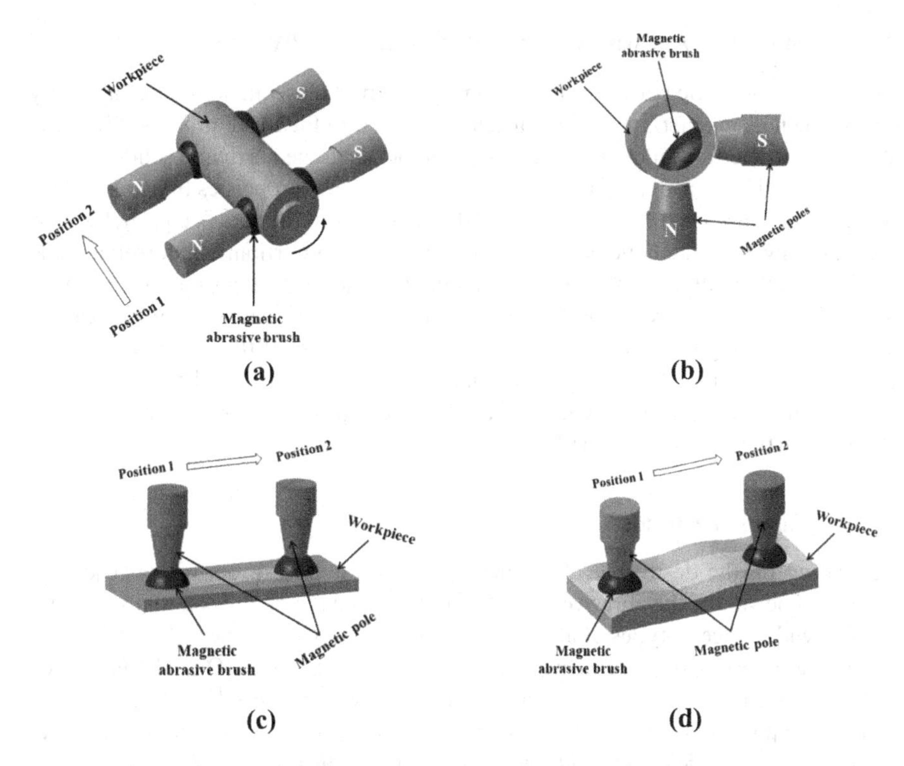

FIGURE 2.1 Different finishing tool configurations, poles settings, and finishing methods: (a) external cylindrical finishing, (b) internal cylindrical finishing, (c) finishing of flat surface, (d) finishing of freeform surface.

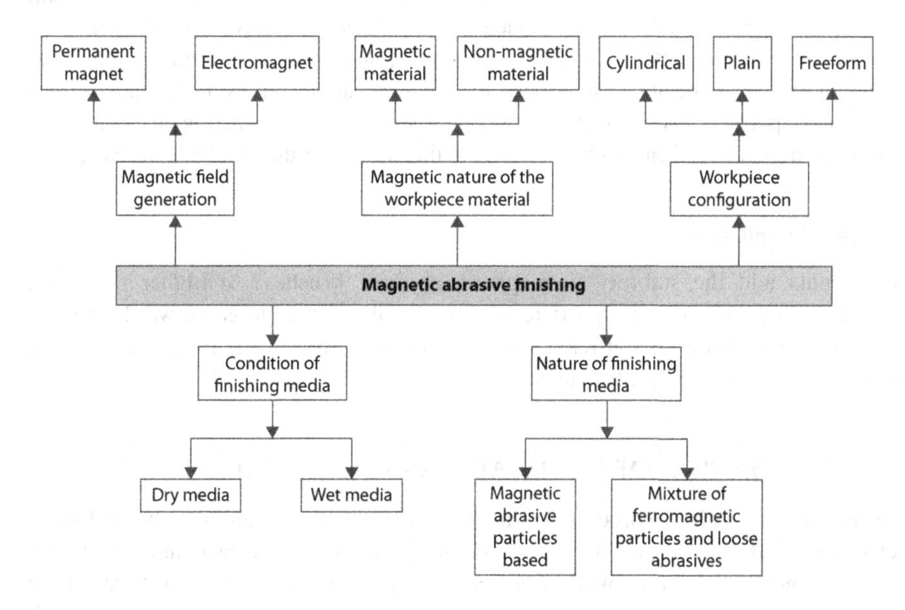

FIGURE 2.2 Classification of magnetic abrasive finishing process.

2.2.2 Magnetic Abrasive Particles/Ferromagnetic Particle

Due to high magnetic permeability, ferromagnetic particles with abrading capability (also known as magnetic abrasive particles) are used in the MAF process. Magnetic abrasive particles are filled in the working gap between the north and south poles of a magnet. Either ferromagnetic particles having abrading properties (magnetic abrasive particles) are used or, in absence of these, a mixture of abrasive particles with non-abrading ferromagnetic particles is used in some proportion. The ferromagnetic particles along with loose abrasives (or magnetic abrasive particles), in the presence of magnetic lines of forces, form the flexible magnetic abrasive brush, which can be used as a finishing tool for finishing the surfaces. The size of the magnetic abrasive particles varies from 50 to 400 μm. The magnetic flux density, and hence the finishing intensity, depends on the type, size, and magnetic properties of the ferromagnetic abrasive particles used in the MAF process.

2.2.3 Abrasive Particles

Depending on the desired level of surface finish, the size of abrasive particles is decided. The size of the abrasive particle varies from nanometer level to micrometer level. A wide range of types of abrasives are available based on the type and harness of material to be finished. Abrasives such as aluminum oxide (Al_2O_3), silicon carbide (SiC), boron carbide, cubic boron nitride, or diamond can be used in MAF.

In MAF process, both bounded and unbounded abrasives are used. In bonded abrasives, abrasives particles are held with ferromagnetic particles by some suitable process, such as sintering [16]. This complete abrasive, along with ferromagnetic particle, is known as a magnetic abrasive particle. In case of bonded abrasive particles, the abrasive particles do not separate out from the ferromagnetic particles; hence, finishing can be done at higher speeds. In unbounded abrasives, loose abrasives are mechanically mixed with ferromagnetic particles of suitable size and used as a mixture of these two components. Since the abrasive particles tend to be separated out from the chains of ferromagnetic particles at higher speeds due to the centrifugal forces acting on the abrasive particles, higher speed cannot be achieved in finishing with unbounded abrasives.

2.2.4 Lubricants

Lubricants add the stability to magnetic abrasive brushes. At higher rotational speeds, they tie the abrasive particles and do not allow them to go outward from the working zone. They also provide some sort of flexibility to the magnetic abrasive brush in finishing of complex surfaces.

2.3 MECHANISM OF MATERIAL REMOVAL IN MAF

While finishing by MAF process, the magnetic abrasive acts under two types of forces. One is normal force (F_N), which is normal to the workpiece surface and responsible for nano indentation in the workpiece surface. This force also keeps together all the ferromagnetic particles in the working gap. The second is tangential force (F_T), which is tangential to the rotation of the workpiece; it is responsible for the chipping at the microscopic level [17]. It is hypothesized that both the forces (normal as well as tangential force) are combined to act on the asperities of the workpiece surface through the magnetic abrasive particle. Due to the contact of abrasive particles with the asperities on the workpiece surface under the action of normal and tangential forces, gradually the asperities are sheared off and a smooth surface is achieved. The process is continued until the desired level of surface finish is achieved.

The role of the magnetic forces acting on ferromagnetic particles can be understood better by the illustration in Figure 2.3. In this figure, a magnetic pole, workpiece, and the cluster of magnetic abrasive particles in the zone of the magnetic field are shown. The finishing mechanism can be understood by taking an example of a ferromagnetic abrasive particle just touching the rotating workpiece at position B.

The normal force (F_N) generated due to the magnetic field gradient will act on the abrasive particle and will penetrate abrasive into the workpiece surface. The tangential force (F_T) will also be generated by the magnetic field present in the working gap. The direction of tangential force (F_T) will be in the same direction as that of workpiece rotation at the location of force application. Due to workpiece rotation, the abrasive particle will experience a resisting force (R_T), which will be opposite to the tangential force (F_T). The resisting cutting force (R_T) is responsible to keep the abrasive particle in its position during the finishing operation. The ferromagnetic particle exerts abrasion forces on the abrasive particle beneath it to remove the material from the workpiece surface. Under the influence of both forces, the abrasive particle removes the work material in the form of microchips and the process continues until the surface unevenness is removed completely.

The forces acting on a single ferromagnetic abrasive particle are depicted in Figure 2.3(b). The same ferromagnetic particle is shown in Figure 2.3(a) at the position A in the magnetic field. The magnetic force (F) is a combination of force

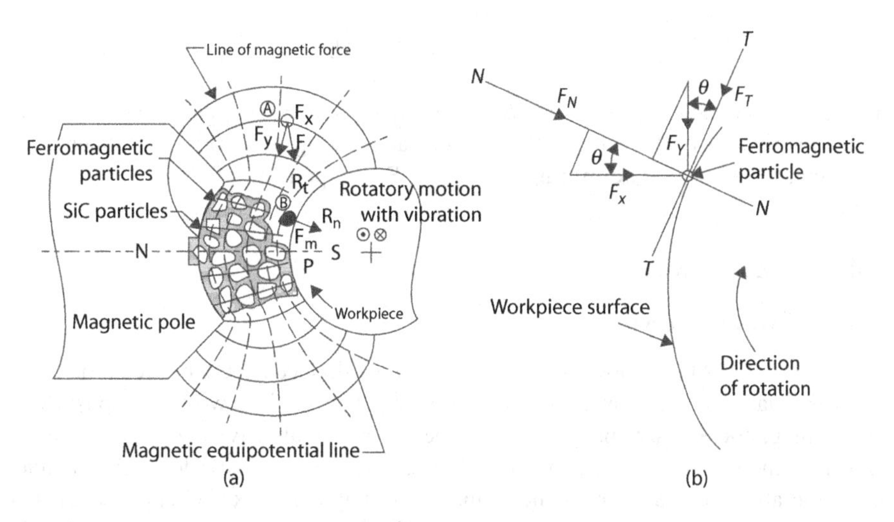

FIGURE 2.3 Schematic view of (a) finishing mechanism in magnetic abrasive finishing [18], (b) forces applied on a single ferromagnetic particle [19].

components F_X and F_Y. The direction of F_X is along the direction of magnetic line of forces and force component F_Y is in the direction of equipotential lines. The F_X component of force pushes the abrasive onto the workpiece surface, due to which the abrasive particle creates micro-indentation in the same. The forces F_X and F_Y can be represented as [20].

$$F_x = \chi_{FP}\, \mu_0 VH\left(\frac{dH}{dX}\right) \tag{2.1}$$

$$F_y = \chi_{FP}\, \mu_0 VH\left(\frac{dH}{dY}\right) \tag{2.2}$$

where

x = direction along the direction of magnetic line of forces
y = direction along the direction of equipotential lines
χ_{FP} = magnetic susceptibility of ferromagnetic particle
μ_0 = magnetic permeability of space
V = volume of ferromagnetic particle
H = magnetic field strength at a specific location (in present case, it is B)
dH/dX and dH/dY = magnetic field gradients along X and Y direction, respectively

The normal (F_N) and tangential force (F_T) along the N-N and T-T direction, respectively, can be represented by [19]:

$$F_N = F_X \cos\theta + F_Y \sin\theta \tag{2.3}$$

$$F_T = -F_X \sin\theta + F_Y \cos\theta \tag{2.4}$$

The angle (θ) depends on the location of the ferromagnetic particle in the magnetic field. On the basis of the value of (θ), the normal and tangential force component can be calculated after predetermining the F_X and F_Y component of forces.

2.4 TYPES OF MAF

2.4.1 CYLINDRICAL MAF

In cylindrical MAF, as shown in Figure 2.4, the workpiece is held between the gaps of unlike magnetic poles by a suitable rotating mechanism. The working gap is filled by the magnetic abrasive particles. Both the choices of abrasive particles, bounded and unbounded, can be used in this process. The abrasive particles form flexible magnetic abrasive brush under the influence of magnetic flux and apply magnetic finishing forces on the cylindrical workpiece. The lap of the magnetic abrasive brush adjusts its compliance according to the surface profile.

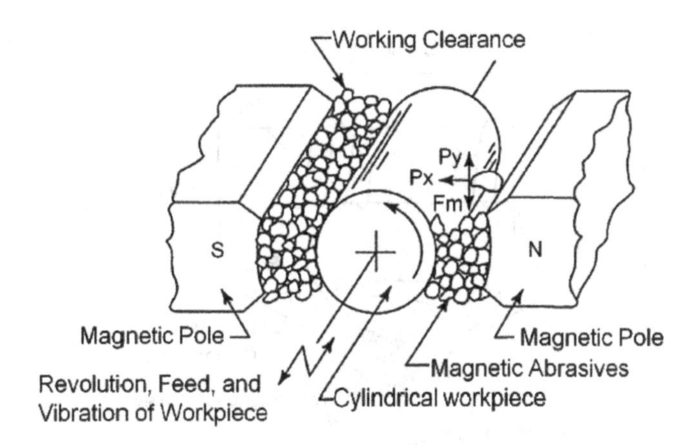

FIGURE 2.4 Cylindrical magnetic abrasive finishing [21].

The finishing is achieved when the rotary motion is provided to the cylindrical workpiece. Simultaneously, the axial movement to the magnetic flux is provided by the vibratory motion to the magnetic poles. Under the vibratory action, the multipoint finishing brush is subjected to change in its shape. Sometimes lubricant is also added with the magnetic abrasive particles to enhance the stability and flexibility of the magnetic abrasive brush.

2.4.2 Internal MAF

An MAF process can also be used for finishing internal surfaces. There are two different approaches used for finishing internal surfaces. In one approach, the magnetic abrasive particles, or a mixture of abrasive particles with ferromagnetic particles, are filled inside the workpiece whose internal surface is to be finished. The workpiece is rotated in between the opposite magnetic poles. In the second approach, a set of opposite magnetic poles is rotated with respect to the workpiece. In both the approaches, the magnetic field generator can be a permanent magnet or the electromagnet. The magnetic abrasive particles align themselves along the magnetic force lines between the opposite poles by the dipole-dipole connection and form a flexible magnetic abrasive brush. Due to the magnetic attraction, these particles, in a cluster form, are pulled out against the workpiece surface by the magnetic poles and create micro-indentation on the workpiece surface. The magnetic tangential force applied on the abrasive particle removes the material in the form of microchips. Figure 2.5 shows a schematic of internal finishing of a tube using the MAF technique.

2.4.3 Plane MAF

Plane MAF is used for finishing of flat, large-size workpieces. The magnetic abrasive particles or a mixture of ferromagnetic particles and abrasive powder is filled in the gap between the magnetic pole and the workpiece. The magnetic field is generated across the working gap by the magnetic pole as shown in Figure 2.6. A small gap of 1–3 mm is maintained for the magnetic abrasive brush formation. The ferromagnetic

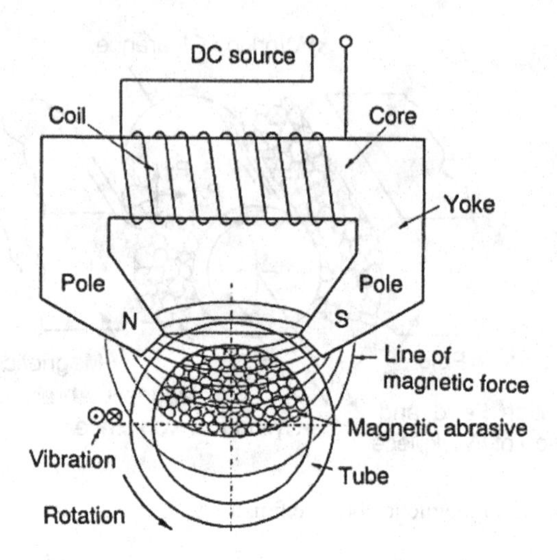

FIGURE 2.5 Internal finishing of a tube by magnetic abrasive finishing [22].

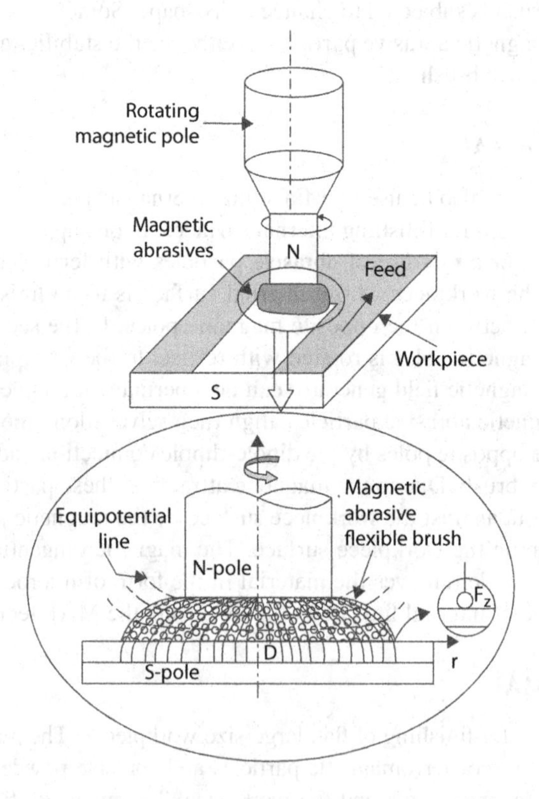

FIGURE 2.6 Plane magnetic abrasive finishing [23].

particles arrange themselves along the magnetic lines of forces by the dipole interaction. The magnetic lines emerge from one magnetic pole (N pole), travel through the working gap, and terminate at the other magnetic pole (S pole). The normal force generated because of the magnetic field gradient is responsible for the penetration of abrasive particles in the workpiece surface, and the rotational movement of the magnetic pole generates tangential force due to which materials wear out in the form of microchips. Normal force is controlled by many factors, such as magnetic field intensity, size of ferromagnetic particles, magnetic properties of ferromagnetic particles, percentage composition of ferromagnetic particles in the magnetic abrasive brush, and working gap while tangential force depends on the rotational speed of the magnetic pole. For finishing of a larger area, feed motion is provided to the rotated magnetic pole.

2.5 HYBRID MAF PROCESSES

During the development of MAF process, the researchers observed that hybrid magnetic abrasive processes perform much better than MAF process. Hybrid MAF processes were developed to make the process faster, more efficient, and more precise as compared to a conventional MAF process. In hybridization of MAF, some other machining/finishing techniques have been incorporated with the MAF to add to their respective advantages and to eliminate their shortcomings. Some of the reported hybrid MAF processes are discussed in the following subsections.

2.5.1 Electrolytic MAF

At present industrial demand of high-quality surface finishes along with high-efficiency is difficult to achieve by a single traditional process. Hence, a compound process involving many machining processes has evolved. The electrolytic MAF process is one of those in which electrolysis and traditional MAF have been combined to obtain high quality of surface finish with higher efficiency of the process. In electrolytic MAF process, electrolyte supply is maintained in the gap between the workpiece and the electrode. In this process, the workpiece is connected to the cathode, and the electrode is connected to the anode of the DC power source. The process principle is shown in Figure 2.7(a).

In normal electrolytic process (without experiencing magnetic field), the negatively charged ions move linearly toward the anode surface (workpiece) and form a passive layer. Since the electric field density is higher at the tip of the peak of the surface roughness than the valley, the thickness of the passive layer is gradually higher towards the tip of the peak of the surface roughness and minimum at the valley, as shown in Figure 2.7(b). But when an electrolytic process is carried out in the presence of a magnetic field, the movement of negatively charged particles (electrolytic ions) follow cycloid curves under the influence of a Lorentz force. This cycloid motion of the electrolytic ion enhances the chances of a collision of ions with the unionized electrolyte, which increases the rate of electrochemical reactions of the electrolytic process [13].

Further, it has been observed that in a magnetic field-assisted electrolytic process, the thickness of the passive surface is much higher than the thickness of the passive surface achieved without a magnetic field. The passive surface is softer than the

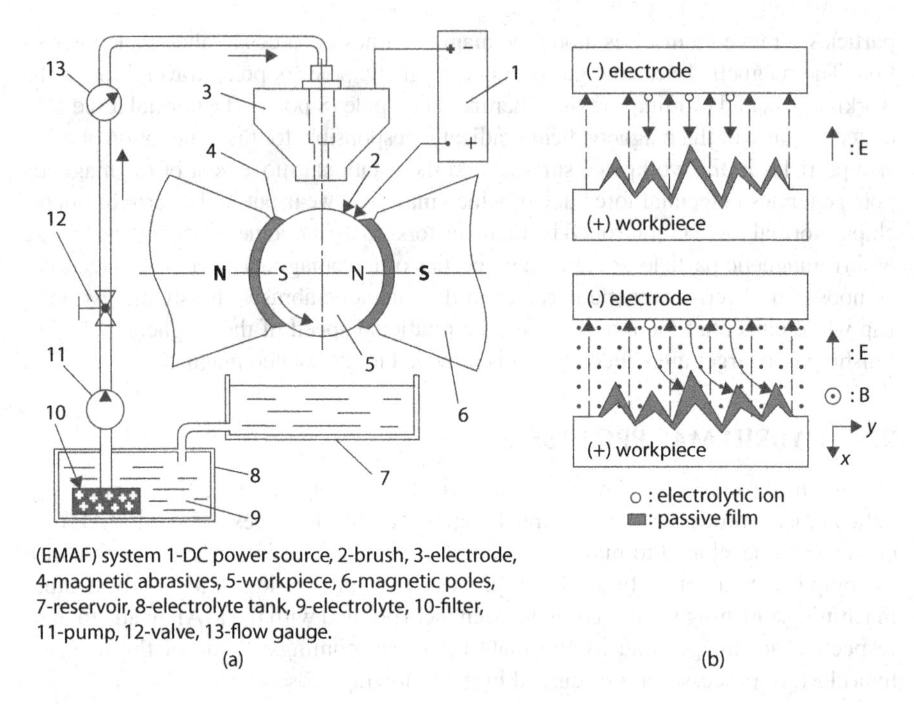

(EMAF) system 1-DC power source, 2-brush, 3-electrode, 4-magnetic abrasives, 5-workpiece, 6-magnetic poles, 7-reservoir, 8-electrolyte tank, 9-electrolyte, 10-filter, 11-pump, 12-valve, 13-flow gauge.
(a)

(b)

FIGURE 2.7 (a) Set-up of magnetic abrasive finishing process, (b) mechanism of formation of passive surface [13].

workpiece material, which can be removed easily in higher volume by the abrasive action of MAF. Thus, the efficiency of the electrolytic MAF becomes higher than the traditional MAF.

2.5.2 Vibration-Assisted MAF

A vibration-assisted MAF process has been developed to improve the finishing performance of traditional MAF. Several researchers have performed various types of studies on this process by providing vibrations either to the tool or to the workpiece. Also, the vibrations are applied in one direction (in *x* or *y* direction to the workpiece or *z* direction to the tool) or in two directions simultaneously (in *x* and *y* direction to the workpiece) or in the horizontal direction (to the workpiece) and vertical direction (to tool) simultaneously [24–27]. It is hypothesized that the applied vibrations enhanced the performance of the MAF process two ways as follows: (i) the applied vibrations enhance the frequency of indentations of the abrasives due to which a single abrasive particle removes more material, (ii), the applied vibrations vary the magnetic field due to which the magnetic field behaves dynamically in nature and applies extra force on the ferromagnetic particles during the finishing operation. Figure 2.8 shows the schematic of vibration-assisted MAF in the deburring process and the shape deformation in the magnetic abrasive brush under the application of vibrations during the process.

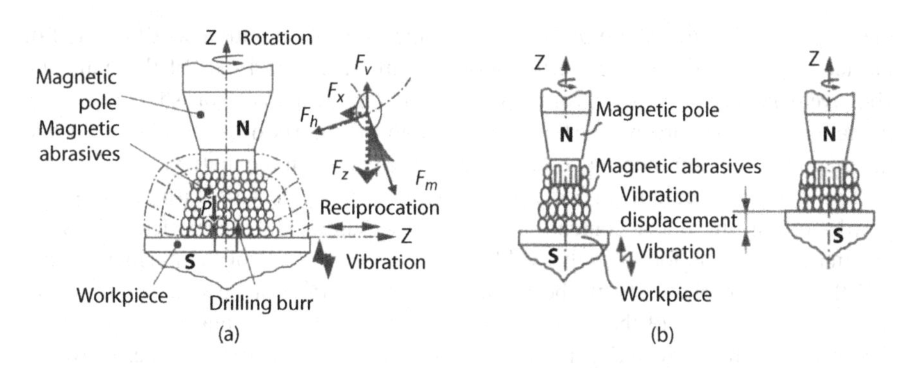

FIGURE 2.8 Vibration-assisted magnetic abrasive finishing. (a) Vibration-assisted deburring process [25], (b) deformation in a magnetic brush [27].

2.6 FACTORS AFFECTING MAF

Generally, the process parameters are categories into input process parameters and output process parameters. Input parameters are the process-affecting parameters and the output parameters are the responses of the input parameters, which are also called response parameters. In MAF, output or response parameters are surface roughness, material removal rate, and accuracy. For analyzing the performance of a process, one or more response parameters are considered simultaneously. The knowledge of process parameters of the MAF process is essential for controlling the performance and to optimizing the process to get maximum output with minimum input. There are several process parameters that affect the performance of the MAF process. Some of the important input parameters are as follows:

i. Magnetic abrasive type and composition
ii. Abrasive particle size
iii. Magnetic flux density
iv. Magnetic device
v. Working gap
vi. Rotational speed
vii. Grinding oil
viii. Workpiece material
ix. Axial vibration

The effects of abovementioned input parameters on MAF performance are discussed in the following subsections.

2.6.1 Magnetic Abrasive Type and its Composition

In MAF, magnetic abrasive particles can be used in bounded or unbounded forms. In bounded forms, the abrasive phase and the magnetic medium phase (ferromagnetic particles) are sintered together but in unbounded form, and the abrasive phase and the magnetic-medium phase are mixed in some proportion with a small amount of lubricating oil. In the first case, the abrasives are an integral part of the magnetic abrasive particle, and hence it cannot leave the magnetic abrasive brush. However, in the latter case, the abrasive particles are loose in the matrix of abrasive powder and the magnetic-medium phase and hence are free to move in the matrix.

The magnetic medium is generally comprised of ferromagnetic particles such as a powder of pure iron. The content of magnetic medium in the matrix determines the amount of magnetic forces in the MAF process. The contents of the magnetic phase and the abrasive phase should be optimum in the matrix for best finishing performance. If the content of the magnetic phase is very high in the matrix, the finishing forces will be more, resulting in an increase in the process performance. However, due to higher forces, surface roughness will increase. If the abrasive content in the matrix is kept high compared to the magnetic phase, the finishing time will be more due to less finishing forces, but the surface roughness will improve due to the action of more abrasive particles. The effect of magnetic forces can be understood by the simple example that if the content of a magnetic phase become very low in the matrix, then due to very low magnetic forces on the abrasive particles, the abrasive particles will fly during the finishing and negligible finishing effect will be observed during the process.

A study on the effect of bounded and unbounded magnetic abrasive particles [16] showed that unbounded abrasive particles give a higher metal removal rate compared to bonded abrasive particles. The higher removal rate could be attributed from the fact that in the case of unbounded abrasives, the abrasive particles are free to move and can scratch much deeper as compared to in bounded cases. In this study in ref. [16], the researcher reported that with unbounded abrasive particles, a 15–20 times higher material rate was observed as compared to that with bounded abrasive particles. In their study, the researchers suggested using unbounded abrasive particles if the initial finish is rough, as in fine-tuning. But if the initial finishing of the product is in the semi-finish category, such as achieved in grinding, the use of bounded abrasive particles is recommended.

2.6.2 Abrasive Particle Size

In bonded magnetic abrasive particles, the abrasive particles are sintered on the surface of the ferromagnetic particles, as shown in Figure 2.9. Let us assume that the diameter of the ferromagnetic particle is D, the diameter of the abrasive particle is d, and the number of abrasive particles attached to a single ferromagnetic particle is n.

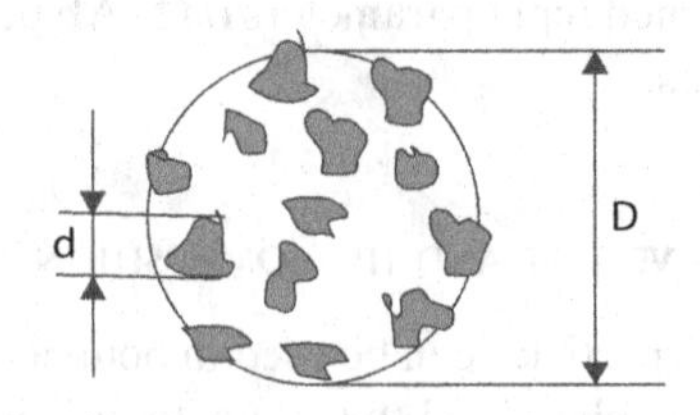

FIGURE 2.9 Schematic of magnetic abrasive particles.

2.6.2.1 Effect of Abrasive Particle Size on Surface Roughness

In MAF, a study on the effect of magnetic abrasive particle size and abrasive particle size (in bonded form) shows that the finishing force, f, applied by a magnetic abrasive particle depends on the size of the magnetic abrasive particle (D) and can be expressed as per equation 2.5 [9].

$$f = k_1 D^{\alpha} \quad \left(\alpha = 2 \text{ or } 3,\ k_1 = \text{constant}\right) \tag{2.5}$$

Assuming that all the abrasive particles are working simultaneously, the force applied by a single abrasive particle (Δf) can be expressed by equation 2.6.

$$\Delta f = \frac{k_1 D^{\alpha}}{n} \tag{2.6}$$

Since the surface roughness is a function of finishing force (as the finishing force increases, the surface roughness increases), so surface roughness will increase as the Δf increases. From equation 2.6, the finishing force per abrasive particle increases as the diameter of the magnetic abrasive particle D increases, and the value of n decreases. Also, the value of n depends on the abrasive particle diameter d; as d increases, the value of n decreases and vice versa.

So, the value of finishing force per abrasive particle (Δf), and hence the surface roughness, increases as the diameter of magnetic particle (D) increases, and the diameter of abrasive particle (d) also increases (i.e., n decreases). From this analysis, we can conclude that to achieve a smooth surface finish, the diameter of the magnetic abrasive particle (D), as well as the diameter of abrasive particle (d), must be smaller.

2.6.2.2 Effect of Abrasive Particle Size on Material Removal

The stoke removal by an abrasive particle (m) can be expressed by equation 2.7 [9].

$$m = k_2 D^{(\alpha\beta - 2)} n^{(1-\beta)} \quad (k_2 = \text{constant}) \tag{2.7}$$

The value β is influenced by the shape of the cutting edge of abrasive particle and its value if 1 and 1.5 for cone- and spherical-shaped cutting edges, respectively.

Assuming that the shape of the cutting edge is a cone, equation 2.7 can be expressed as:

$$m = k_2 D^{(\alpha - 2)} \tag{2.8}$$

Therefore, it can be seen from equation 2.8 that the stock removal (m) depends on magnetic particles diameter (D) and is not affected by the diameter of abrasive particle (d).

On the other hand, assuming that the cutting edge is a sphere, equation 2.7 can be expressed as:

$$m = k_2 D^{(1.5\alpha - 2)} n^{(-0.5)} \tag{2.9}$$

As a result, the stock removal (m) increases as the diameter of the magnetic abrasive increases and the value of n decreases (i.e., diameter of abrasive d increases).

From these analyses, we observe that to achieve a higher finishing rate as well as smooth surface finish, the diameter of the ferromagnetic particles must be selected as a compromise between these two desired parameters. The smaller-diameter abrasive particles are good for better surface finish.

2.6.3 Magnetic Flux Density

In MAF, the magnetic field is either generated by employing the permanent magnet or by the electromagnet. When the external magnetic field is applied, all the magnetic domains in the magnetic phase arrange themselves in the direction of the applied magnetic field due to which the magnetic substances magnetize. From the B-H curve of any magnetic substance, it is evident that there is a magnetic saturation level beyond which the magnetic substance cannot be magnetized even after further increasing the applied magnetic flux to higher levels. The magnetic flux density of magnetic abrasive particles can be enhanced by the following two methods: (1) by increasing the proportion of magnetic phase in the magnetic abrasive particles, or (2) by selecting a magnetic phase that has higher magnetic saturation intensity. The magnetic flux density is measured in Gauss or in Tesla (T) and is expressed as:

$$B = \mu_o \mu_r H \tag{2.10}$$

where:

B = magnetic flux density (T)
μ_o = magnetic permeability in vacuum, ($\mu_o = 4\pi \times 10^{-7}$ H/m)
μ_r = relative magnetic permeability
H = magnetic field strength (A/m)

Increasing the magnetic field (or magnetizing current in the case of an electromagnet) increases the magnetizing force and hence the finishing forces on the magnetic abrasive particles. In one of the studies on the effect of magnetic flux density on the surface roughness and material removal rate [16], researchers reported that an increase in magnetic flux density increases the surface finish and the material removal rate up to a certain level beyond which it reaches a saturation value. They [16] observed these improvements when they changed the magnetic flux density from 0.17 to 0.37 T. They also reported that with unbounded abrasive particles, better surface finish (Ra less than 50 nm) and higher removal rate was observed at 2 A current density, which corresponded to a magnetic flux density of 0.37 T.

The selection of magnetic flux density in the MAF process depends on many factors, such as workpiece material, finishing requirement, and time of finishing, etc. Generally, most of the finishing work can be carried out with a magnetic flux density ranging from 0.6 to 1.4 T.

2.6.4 Working Gap

Magnetic flux density depends on the working gap. As the working gap increases, the magnetic flux density decreases and vice versa. The magnetic forces and hence the finishing forces, which are responsible for finishing, are a function of magnetic flux density. So, as the working gap increases or decreases, the magnetic force (also the finishing force) decreases or increases, respectively. From experimental studies, it has been observed that a decrease in working gap causes higher finishing forces due to which the material removal increases, and better surface finish is achieved [21]. Due to the higher magnetic forces at the lower working gap, the magnetic abrasive brush becomes stronger and applies deeper cuts while finishing. Contrary to this, at a higher working gap, the segregation of magnetic particles and abrasives take place, i.e., all the magnetic particles concentrate near the magnet and abrasives near the workpiece surface. Due to the segregation of magnetic particles and abrasives, the abrasive particles fall apart from the machining zone during finishing and the problem becomes severe at higher finishing speeds. At a lower working gap, once the abrasive particles become blunt, it is less effective to replenish the abrasive particles in the finishing media compared to replenishing abrasives in a higher working gap. Many researchers found that, in the MAF process, the working gap is a crucial and most significant process parameter [21, 28, 29].

2.6.5 Rotational Speed

The rotational speed is required to provide the relative motion between the workpiece and the MAF brush due to which the workpiece material is removed in the form of microscopic chips. Researchers have performed experiments at different rotational speeds to know the behavior of rotational speed in MAF process.

The researchers have reported that the material removal and the percentage improvement in surface finish (% ΔRa) increases on increasing the rotational speed. But at higher rotational speeds, the abrasive particles scatter under the influence of higher centrifugal forces on the abrasive particles, and it causes a decrement in material removal, as well as in percentage improvement in surface finish (% ΔRa) after an optimum value [21, 30].

2.6.6 Axial Vibration

The axial vibration to the magnetic abrasive brush through the magnetic poles causes the stirring action in the magnetic abrasive brush and pushes the new abrasives (or the cutting edges) towards the workpiece surface. During finishing under axial vibration, every time fresh abrasive particles participate in finishing action, this results in higher material removal and better surface finish.

In one of the studies on vibration-assisted MAF, the researchers reported that with a vibration of 15 Hz and amplitude of 2 mm the improvement in material removal reaches as much as 1.5 times more than without vibration [31].

In another study on vibration-assisted MAF, the workpiece was rotated at a speed of 3000 rpm, and the magnetic poles were subjected to vibrations of frequencies from 0 to 12 Hz. The vibrations were generated using an electromotive slider controller. The experiments were performed on an STS 307 of length and diameter 60 mm and 3 mm, respectively. In their study, they reported that at 0 Hz vibration, initially the surface finish got some improvement but later at around 60 s of finishing, the surface finish became worse than the initial surface finish of the workpiece. In this process, the best surface finish achieved was 0.51 μm. On the other hand, when the magnetic poles were subjected to 12 Hz vibration, the surface finish improved gradually until 30 s of finishing, and after that the roughness stabilized till further machining [32].

To increase the removal rate at a higher rotational speed of the workpiece, both the frequency of magnetic pole vibration and the amplitude of vibration needs to increase [16].

2.6.7 Workpiece Material

The mechanical and magnetic properties such as hardness, strength, magnetic, non-magnetic, etc. of the workpiece material are critical and affect the performance of the MAF process in terms of surface roughness and removal rate.

In one of the studies, workpieces of hardness HRC61 and HRC55 were finished using an MAF process to analyze the effect of workpiece hardness on surface finish and material removal [18]. They conducted experiments on three types of magnetic abrasive particles, namely: steel grit alone (size 180 μm), a mixture of steel grit and SiC abrasive particles of size 1.2 μm, and a mixture of steel grits and SiC abrasive particles of size 5.5 μm. In their studies, they reported that for workpieces having hardness HRC61, the highest material removal and the best surface finishing were achieved with magnetic abrasive particles having a mixture of steel grits and SiC abrasive particles of size 5.5 μm in 25 min of finishing. In comparison to finishing results (surface finish and material removal) on a workpiece of HRC55 in 25 min of finishing, they observed that surface finish and material removal was higher for workpiece having higher hardness (HRC61) as compared to workpiece having lower hardness (HRC55). The reason behind this was explained as: the abrasive particles in workpiece of lower hardness penetrates deeper and face higher resistance as compared to workpiece having higher hardness. Due to higher resistance, the abrasive particles roll out without performing any cutting. The abrasive particles, in finishing of harder materials, penetrate marginally and hence shallow penetration takes place due to which the abrasive experience lowers resistance forces. Under the influence of lower resistance forces, the abrasive particles do not roll out, but they slide, due to which the material removal increases and the surface roughness decreases [18].

The finishing results on copper (c 61400) and the stainless steel (SS 202) alloy using a double disc MAF process was reported by Kala and Pandey [33]. In their studies, they found that the best surface finish achieved on copper and stainless-steel alloys was 53 nm and 79 nm, respectively. Nteziyaremye et al. [34] performed the experiments on finishing of internal and external surfaces of stainless steel 316 needles, which are used in breast cancer biopsy operations. They reported that the surface roughness improved from 0.4–0.5 μm to 0.01 μm in 5 min of finishing.

Gao et al. [35] performed experimental analysis on finishing of paramagnetic materials (copper 27400 and stainless steel 316) using atomized magnetic abrasive powder. In their studies, they reported that the material removal was higher for copper than stainless-steel, but the surface finish achieved on copper workpieces was rougher than the minimum surface roughness recorded for stainless steel workpieces.

2.6.8 Cutting Fluids (Lubricants)

Cutting fluids or lubricants reduces the cutting forces and the frictional forces between the abrasive and the workpiece surface. In the MAF process, the cutting fluid helps make better contact between abrasive particles and the magnetic phase by filling the voids or gaps between them. It also helps to avoid the separation between the abrasive and magnetic phases at high finishing speed and acts the same as magnetic fields do in the MAF process (acts as a binder). In MAF, the cutting fluid or lubricant also controls the temperature generation in the finishing zone and maintains it at a minimum level.

Goloskov et al. [36] found that by simply changing the composition of the cutting fluid, the productivity of the process can be increased six to seven times without altering the other process conditions. Cutting fluids are sometimes used to generate a passive layer on the workpiece surface by a chemical reaction. The mechanically weak passive layer is easy to remove with less cutting forces; hence, it is easy to finish hard materials, as well as increasing the material removal significantly. Surface active fluids also reduce the cutting forces in microcutting processes by destroying the hard surface layer of the component. Emulsol E2 (soluble oil) is one of the cutting fluids that imparted good results when used at 5%–10% in the solution.

Fox et al. [16] employed zinc stearate as the solid lubricant in the MAF for studying the effect of lubrication on the surface finish. They reported that the lubrication increases the flexibility of the magnetic abrasive brush and hence increases the ability of the brush to impart better surface finish in the process. They also reported that at up to 5 wt% lubricant, the surface finish of the component increases during the finishing. A particular surface finish can be achieved in lesser time if the lubricant or the cutting fluid is used in the process.

2.6.9 Finishing Time

Selecting a finishing time is also a crucial factor in the MAF process. If all other factors are selected properly, it has been seen that the surface finish and the material removal increase on increasing the finishing time [9, 16, 37]. With all the optimum parameters, only a critical surface finish can be obtained, beyond which the finishing is not possible. To achieve surface finish beyond the critical value, the process parameters need to be adjusted, such as reducing the size of the magnetic abrasive particle, making the finishing condition gentle by reducing the magnetic flux density, etc. Without altering the process parameters if the finishing operation is continued, either the surface finish remains constant or deteriorates due to unnecessary rubbing and scratching of the surface by the abrasive particles.

2.7 ADVANTAGES OF MAF

With the MAF process, it is possible to get a mirror-like surface finish irrespective of the surface being either an internal or external surface. The following advantages make this process unique over other alternative finishing processes:

- As the process involves a low level of finishing forces along with loose particles of magnetic abrasives, the damage to the surface is negligible.
- Simultaneous finishing of similar parts, finishing of combined surface such as grooves and cylindrical surfaces, cylindrical and conical surfaces, etc., are possible.
- Since it does not require costly machines and tooling, the process is economical.
- The tool, which is magnetic abrasive brush, is self-sharpening and does not require any dressing and compensation.
- The tool is self-adaptable so that it can be used for finishing complex geometries of the workpiece.
- The finishing forces, and hence the finishing action, can be controlled easily if an electromagnet is employed in the process for the magnetic field generation.
- The process is most suitable for brittle materials, such as glass and ceramics, because it does not generate defects such as microcracks on the surface.
- The finished surface is free from burrs and thermal damages.
- The process is environmentally friendly due to no contamination in the environment during and after the process, no toxic byproducts etc.
- The physical and mechanical characteristics of the finished product are substantially high, such as long life, wear resistance, corrosion resistance, high load-bearing capacity, etc.
- Soft materials such as copper, aluminum, brass, and their alloys can also be finished precisely and accurately with sufficient ease.

2.8 LIMITATIONS OF MAF

Like any other finishing process, the MAF process also suffers from some limitation. Some major limitations are as follows:

- Finishing a microscaled thick material by an MAF process is difficult because the generated magnetic force can damage the workpiece.
- Only a marginal surface finished is possible on ferromagnetic materials such as cobalt and nickel alloys due to the magnetization of work material under the influence of a magnetic field. After magnetization, the work material attracts the magnetic abrasive particles; thus the abrasive particles lose their cutting ability.

- The most significant limitation of the MAF process is its low efficiency and material removal rate. The problem is acute in finishing of hard materials.
- It has been reported that sometimes the magnetic media and abrasives can impregnate workpiece surfaces and deteriorate its physical and optical characteristics, such as color, light reflection, etc.

2.9 APPLICATIONS OF MAF

MAF and its allied processes (different types) have been employed to finish all types of surfaces like planar, non-planar, freeform, internal, and external surfaces of a variety of materials having applications not only in the engineering field but also in the biomedical field. Boggs et al. [38] used vibration-assisted MAF to create a textured mold surface which was replicated onto the silicone leaflets. This polymeric heart valve leaflet having a textured surface helped reduce blood cell adhesion and aggregation. Yamaguchi and Graziano [39] finished femoral knee components made of cobalt chromium alloy using the MAF process.

Kim et al. [40] finished Ni-Ti stent wire using an ultra-precision MAF technique with a rotating magnetic field as shown in Figure 2.10.

They studied and compared the role of different processing oils on the surface finish of the stent wire. The use of environmentally friendly vegetable oil like olive oil and castor oil was compared with the use of industrial light oil commonly used in MAF processes. From the results they obtained (Figure 2.11), they concluded

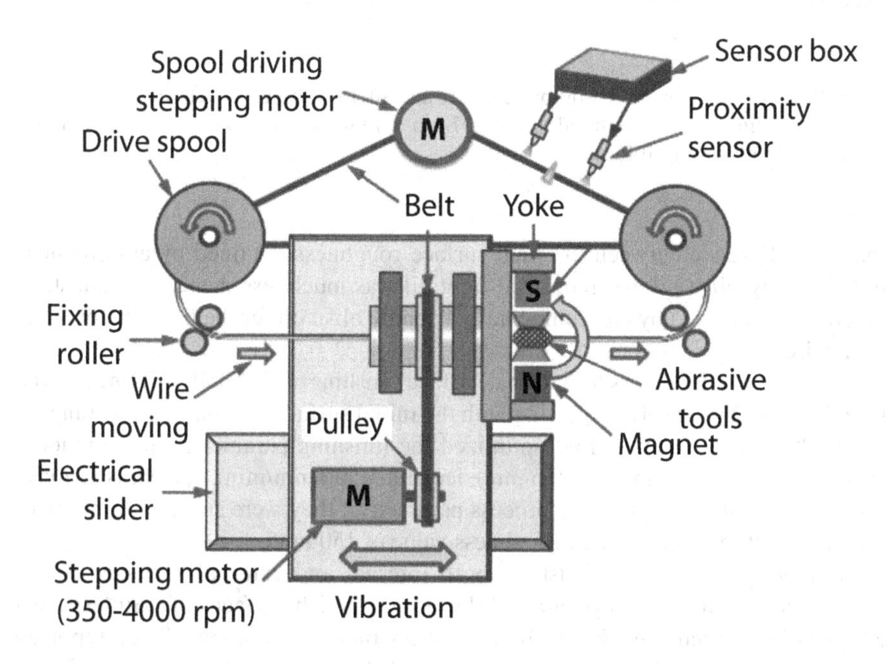

FIGURE 2.10 Schematic diagram ultra-precision magnetic abrasive finishing equipment for finishing of Ni-Ti wire [40].

FIGURE 2.11 Surface of the workpiece before and after finishing. (a) Before finishing, Ra = 0.24 µm; (b) finished with light oil, Ra = 0.07 µm; (c) finished with olive oil Ra = 0.08 µm; (d) finished with castor oil Ra = 0.12 µm [40].

that the difference between the final surface roughness obtained by environmentally friendly olive oil and industrial light oil was much less, hence the industrial oil could be replaced by environmentally friendly olive oil for the finishing of Ni-Ti stent wires.

Using the MAF process for internal surface finishing, Kajal et al. [41] finished the barrel of a 0.32 in. revolver. Starting with the initial surface roughness value ranging from 600 to 900 nm, they first optimized the finishing parameters like rotational speed (282 rpm), working gap (0.6 mm), feed rate (500 mm/min), and abrasive mesh size (800). Using the optimized process parameters, they were successful in finishing the barrel to a final surface roughness value of 150 nm.

In another example of finishing both internal and external surfaces by the MAF process, Nteziyaremye et al. [34] used it to finish surfaces of stainless steel 316 needles, which are used in breast cancer biopsy operations. They reported that the surface roughness improved from 0.4–0.5 um Sa to .01 um Sa in 5 min of finishing.

FIGURE 2.12 Photographs of workpiece before and after finishing [42].

Jiao et al. [42] used the MAF process to finish a seal-ring groove surface, where they deployed this process to simultaneously finish the bottom and sides of the ring groove. The final surface roughness values of the bottom and sides were close to Ra = 0.6 μm, and this value was achieved from an initial surface roughness of Ra = 4.3 μm (Figure 2.12).

REFERENCES

1. Baron, Y. M. (1975). Technology of abrasive machining in a magnetic field. Mashinostroenie, Leningrad, 43.
2. Konovalov, E. G., & Sakulevich, F. J. (1974). Principles of Electro-Ferromagnetic Machining. Nauka I technika, Minsk (in Russian).
3. Makedonski, B. G. (1974). Schleifen im Magnetfeld. Fertigungstechnik und Betried, 24, 230.
4. Sakulevich, F. J., & Kozuro, L. M. (1978). Magneto-Abrasive Machining. Nauka I technika, Minsk (in Russian).
5. Sakulevich, F. J., & Kozuro, L. M. (1977). Magneto-Abrasive Machining of Fine Parts. Vyssaja Skola, Minsk (in Russian).
6. Umehara, N & Kato, K. (1990). Principles of magnetic fluid grinding of ceramic balls. Applied Electromagnetics in Materials, 1, 37–43.
7. Anzai, M., Kawashima, E., Otaki, H., & Nakagawa, T. (1993, June). Magnetic abrasive finishing of WC-CO curved surfaces. Proceedings of the International Conference on Machining of Advanced Materials (pp. 415–422). National Institute of Standards & Technology.
8. Shinmura, T., Yamaguchi, H., & Aizawa, T. (1993). A new internal finishing process of non-ferromagnetic tubing by the application of a magnetic field: the development of a unit finishing apparatus using permanent magnets. International Journal of the Japan Society for Precision Engineering, 27(2), 132–137.
9. Shinmura, T., Takazawa, K., Hatano, E., Matsunaga, M., & Matsuo, T. (1990). Study on magnetic abrasive finishing. CIRP Annals, 39(1), 325–328.

10. Takazazawa, K., Shinmura, T., & Hatano, E. (1985). Development of magnetic abrasive finishing and its equipment. In Proceedings of Deburring and Surface Conditioning, SME Conference, Orlando, USA.
11. Shinmura, T., Hatano, E., & Takazawa, K. (1986). Development of spindle-finish type finishing apparatus and its finishing performance using a magnetic abrasive machining process. Bulletin of the Japan Society of Precision Engineering, 20(2), 79–84.
12. Yamaguchi, H., Shinmura, T., & Kaneko, T. (1996). Development of a new internal finishing process applying magnetic abrasive finishing by use of pole rotation system. International Journal of the Japan Society for Precision Engineering, 30(4), 317–322.
13. Yan, B. H., Chang, G. W., Cheng, T. J., & Hsu, R. T. (2003). Electrolytic magnetic abrasive finishing. International Journal of Machine Tools and Manufacture, 43(13), 1355–1366.
14. Yang, S., & Li, W. (2018). Surface quality and finishing technology. Surface Finishing Theory and New Technology (pp. 1–64). Springer, Berlin, Heidelberg.
15. Komanduri R (1998). Magnetic field assisted finishing of ceramics—Part III: on the thermal aspects of magnetic abrasive finishing (MAF) of ceramic rollers. Journal of Tribology, 120(4), 660–667.
16. Fox, M., Agrawal, K., Shinmura, T., & Komanduri, R. (1994). Magnetic abrasive finishing of rollers. CIRP Annals, 43(1), 181–184.
17. Singh, D. K., Jain, V. K., & Raghuram, V. (2006). Experimental investigations into forces acting during a magnetic abrasive finishing process. The International Journal of Advanced Manufacturing Technology, 30(7–8), 652–662.
18. Chang, G. W., Yan, B. H., & Hsu, R. T. (2002). Study on cylindrical magnetic abrasive finishing using unbonded magnetic abrasives. International Journal of Machine Tools and Manufacture, 42(5), 575–583.
19. Judal, K. B., & Yadava, V. (2013). Modeling and simulation of cylindrical electrochemical magnetic abrasive machining of AISI-420 magnetic steel. Journal of Materials Processing Technology, 213(12), 2089–2100.
20. Smolkin, M. R., & Smolkin, R. D. (2006). Calculation and analysis of the magnetic force acting on a particle in the magnetic field of separator. Analysis of the equations used in the magnetic methods of separation. IEEE Transactions on Magnetics, 42(11), 3682–3693.
21. Jain, V. K., Kumar, P., Behera, P. K., & Jayswal, S. C. (2001). Effect of working gap and circumferential speed on the performance of magnetic abrasive finishing process. Wear, 250(1–12), 384–390.
22. Yamaguchi, H., & Shinmura, T. (1999). Study of the surface modification resulting from an internal magnetic abrasive finishing process. Wear, 225, 246–255.
23. Kanish, T. C., Narayanan, S., Kuppan, P., & Ashok, S. D. (2018). Investigations on wear behavior of magnetic field assisted abrasive finished SS316L material. Materials Today: Proceedings, 5(5), 12734–12743.
24. Natsume, M., Shinmura, T., & Sakaguchi, K. (1998). Study of magnetic abrasive machining by use of work vibration system (characteristics of plane finishing and application to the inside finishing of groove). Transactions of the Japan. Society of Mechanical Engineers, 64(627), 4447–4452.
25. Yin, S., & Shinmura, T. (2004). Vertical vibration-assisted magnetic abrasive finishing and deburring for magnesium alloy. International Journal of Machine Tools and Manufacture, 44(12–13), 1297–1303.
26. Lee, Y. H., Wu, K. L., Jhou, J. H., Tsai, Y. H., & Yan, B. H. (2013). Two-dimensional vibration-assisted magnetic abrasive finishing of stainless steel SUS304. The International Journal of Advanced Manufacturing Technology, 69(9–12), 2723–2733.

27. Yin, S., & Shinmura, T. (2004). A comparative study: polishing characteristics and its mechanisms of three vibration modes in vibration-assisted magnetic abrasive polishing. International Journal of Machine Tools and Manufacture, 44(4), 383–390.
28. Singh, D. K., Jain, V. K., & Raghuram, V. (2004). Parametric study of magnetic abrasive finishing process. Journal of Materials Processing Technology, 149(1–3), 22–29.
29. Givi, M., Tehrani, A. F., & Mohammadi, A. (2010, June). Statistical analysis of magnetic abrasive finishing (MAF) on surface roughness. AIP Conference Proceedings (Vol. 1252, No. 1, pp. 1160–1167). American Institute of Physics.
30. Deepak, B., Walia, R. S., & Suri, N. M. (2012). Effect of rotational motion on the flat work piece magnetic abrasive finishing. International Journal of Surface Engineering and Materials Technology, 2, 50–54.
31. Yin, S. H., Wang, Y., Shinmura, T., Zhu, Y. J., & Chen, F. J. (2008). Material removal mechanism in vibration-assisted magnetic abrasive finishing. Advanced Materials Research (Vol. 53, pp. 57–63). Trans Tech Publications Ltd.
32. Im, I. T., Mun, S. D., & Oh, S. M. (2009). Micro machining of an STS 304 bar by magnetic abrasive finishing. Journal of Mechanical Science and Technology, 23(7), 1982–1988.
33. Kala, P., & Pandey, P. M. (2015). Comparison of finishing characteristics of two paramagnetic materials using double disc magnetic abrasive finishing. Journal of Manufacturing Processes, 17, 63–77.
34. Nteziyaremye, V., Wang, Y., Li, W., Shih, A., & Yamaguchi, H. (2014). Surface finishing of needles for high-performance biopsy. Procedia Cirp, 14, 48–53.
35. Gao, Y., Zhao, Y., Zhang, G., Zhang, G., & Yin, F. (2019). Polishing of paramagnetic materials using atomized magnetic abrasive powder. Materials and Manufacturing Processes, 34(6), 604–611.
36. Goloskov, E. I., Baron, Y. M., & Deryabin, Y. P. (1970). Polishing external cylindrical surfaces by the magnetic-abrasive method. Chemical and Petroleum Engineering, 6(11), 946–948.
37. Jayswal, S. C., Jain, V. K., & Dixit, P. M. (2005). Magnetic abrasive finishing process—a parametric analysis. Journal of Advanced Manufacturing Systems, 4(02), 131–150.
38. Boggs, T., Carroll, R., Tran-Son-Tay, R., Yamaguchi, H., Al-Mousily, F., & DeGroff, C. (2014). Blood cell adhesion on a polymeric heart valve leaflet processed using magnetic abrasive finishing. Journal of Medical Devices, 8(1), 011005.
39. Yamaguchi, H., & Graziano, A. A. (2014). Surface finishing of cobalt chromium alloy femoral knee components. CIRP Annals, 63(1), 309–312.
40. Kim, J. S., Nam, S. S., Heng, L., Kim, B. S., & Mun, S. D. (2020). Effect of environmentally friendly oil on Ni-Ti stent wire using ultraprecision magnetic abrasive finishing. Metals, 10(10), 1309.
41. Kajal, S., Jain, V. K., Ramkumar, J., & Nagdeve, L. (2019). Experimental and theoretical investigations into internal magnetic abrasive finishing of a revolver barrel. The International Journal of Advanced Manufacturing Technology, 100(5), 1105–1122.
42. Jiao, A. Y., Quan, H. J., Li, Z. Z., & Chen, Y. (2016). Study of magnetic abrasive finishing in seal ring groove surface operations. The International Journal of Advanced Manufacturing Technology, 85(5), 1195–1205.

3 Magnetorheological Finishing

3.1 MAGNETORHEOLOGICAL FLUID

The word "magnetorheological" is attributed to the change of rheology of the fluid under the magnetic field. Magnetorheological (MR) fluid is a suspension consisting of a magnetic dispersed phase, a continuous phase, and some additives. The continuous phase is the fluid phase of the MR fluid in which the solid particles are spread about. The continuous phase is commonly known as the carrier medium, as it is responsible for carrying the solid phase in the MR fluid. Mineral oil and water are commonly used as carrier mediums. The role of the magnetic phase is to impart rheological changes to the MR fluid. As the name suggests, the magnetic phase is made up of small micron-sized ferromagnetic materials, the most common being carbonyl iron particle (CIP) and electrolyte iron powder (EIP). The difference between CIP and EIP are in the way they are produced, their shape, and the purity level they possess. Sometimes, additives in the form of chemicals or surfactants are also added to the MR fluid. While chemicals are added to enhance an MR fluid's effective life, surfactants are used to prevent the solid constituents from settling down. A few additives commonly used in MR fluid are glycerol, sodium carbonate, grease, oleic acid, etc. [1].

When abrasive particles are added to the MR fluid, the suspension is termed as magnetorheological polishing (MRP) fluid. This is because with the addition of abrasive to the MR fluid, the fluid can be used for finishing/polishing operations under a controlled magnetic field. Some popular abrasives used in MRP fluid are cerium oxide, diamond powder, aluminum oxide, silicon carbide, and boron carbide [1]. Figure 3.1(a) shows the random distribution of solid constituents of MRP fluid in a non-energized state, i.e., in the absence of a magnetic field. In this state, both magnetic and abrasive particles are distributed all over the carrier medium. The MRP fluid exhibits Newtonian behavior in this non-energized state. However, magnetically energizing the MRP fluid results in the ferromagnetic CIPs forming tiny magnetic dipoles. These tiny dipoles attach to each other along the magnetic field lines, resulting in the formation of a structure that appears to be chain-like. In the event of this chain formation, the non-magnetic abrasive particles are gripped between these chains (Figure 3.1b). The chain-like formation is in the direction of magnetic field lines. The strength of these chains is directly related to the magnitude of strength of the magnetic field, which acts as a binder for CIPs [2]. In a magnetically energized state, non-Newtonian behavior is exhibited by the MRP fluid. Upon removal of the magnetic field, these chains collapse allowing the solid constituents of the MRP to be randomly distributed again.

DOI: 10.1201/9781003228776-3

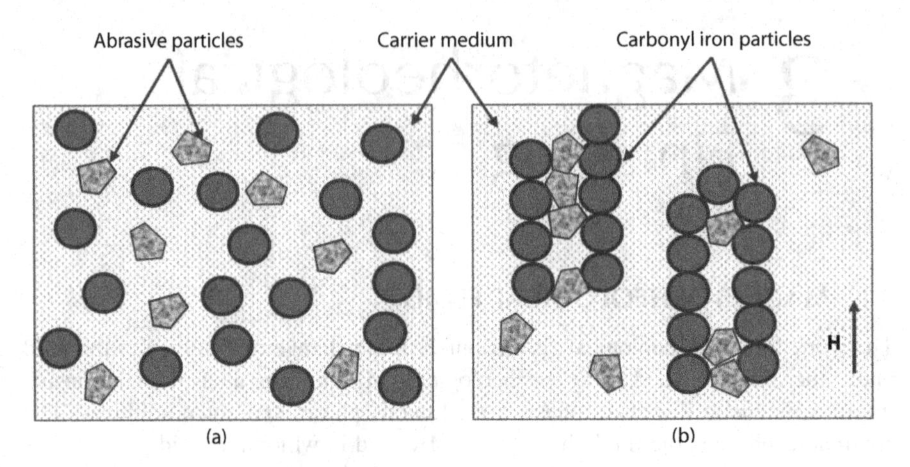

FIGURE 3.1 Magnetorheological effect (a) magnetically energized and (b) non-energized state.

3.2 MAGNETORHEOLOGICAL FINISHING

Magnetorheological finishing (MRF) is one of the magnetic field-assisted finishing processes commonly used for extremely precise surface finishing applications. Since the MRF process relies on MRP fluid, whose shape and stiffness can be magnetically manipulated and controlled in real time, it is counted among the best controllable finishing methods. Also, being a non-direct contact finishing operation (in terms of the interaction between the tool tip and workpiece), the MRF process is considered a preferred choice when it comes to precision finishing of complex shapes without causing any damage to the finished surface. The primary limitations associated with conventional finishing techniques are the uncontrollable finishing forces. In the MRF process, the forces can be controlled, resulting in its ability to meet extreme standards of surface accuracy. In addition to the controllable finishing forces discussed above, the MRF process possesses several other characteristic suitable for finishing applications, such as:

- It has an adjustable compliance that can be altered using the magnetic field.
- It can cool the polishing zone by extracting heat and debris from the finishing area.
- It prevents loading of the polishing tool.
- It can easily adapt and take the shape of the workpiece surface as it is flexible.

The MRF process was initially ideated at Minsk, Belarus. However, its further development as a wholesome process was concluded at University of Rochester's Center for Optics Manufacturing lab, where it was used to automate the finishing of high-precision lenses [3]. In 1998, the MRF process was commercialized by QED Technologies with a wheel-type MRF setup as shown in Figure 3.2 [4]. In the wheel-type MRF system, a continuous supply of MRP fluid is maintained by a fluid

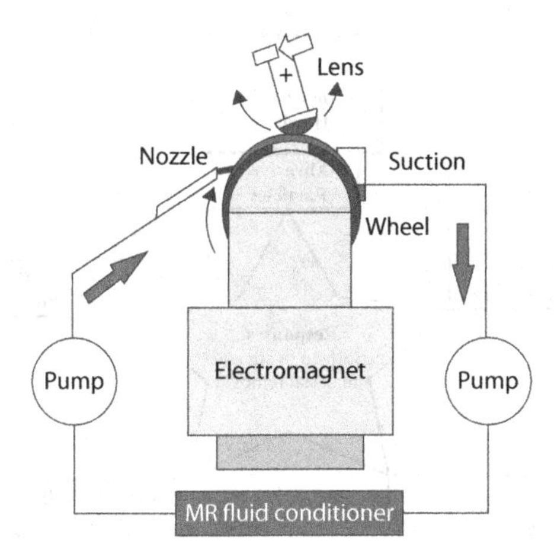

FIGURE 3.2 Schematic diagram of wheel-type magnetorheological finishing set-up (From ref. [4].)

circulatory system consisting of two pumps, a nozzle, and a suction system. The circulatory system is also responsible for maintaining the viscosity of the MRP fluid by altering the quantity of carrier medium in the MRP fluid. Due to the placement of the electromagnet below the wheel, the wheel gets magnetized, and this causes the MRP fluid to form a stiffened ribbon over its periphery. As the wheel is rotated, the stiffened ribbon passes through the gap between the workpiece surface and wheel periphery causing the material removal from the workpiece surface. Swing and rotational motion is given to the workpiece to cover the entire finishing area.

3.2.1 Process Parameters of MRF

Since the MRF process is based on the MRP fluid, the constituents and concentration of MRP fluid plays a major role in influencing the process parameters. The different parameters on which the MRF process depends are shown in Figure 3.3. It is extremely important to control all these parameters simultaneously to churn out the highest yield from the MRF process. The right selection of these parameters plays an important role in getting a desired surface finish on a workpiece. Depending upon the property of the workpiece (composition, hardness, surface roughness), these parameters must be altered to achieve a right combination of parameters to get the job done.

Being a magnetic field-assisted finishing process, MRF draws heavily upon magnetic field strength and quantity and size of ferromagnetic particles. Therefore, these two parameters (magnetic field and CIPs) become more significant among all the other process parameters. This is because in the absence of any one of these, the MRF process becomes incapable of finishing.

3.2.1.1 Magnetic Flux Density

The requirement of an external magnetic field is a must for the MRF process, and this makes the magnetic flux density one of the most prominent process parameters.

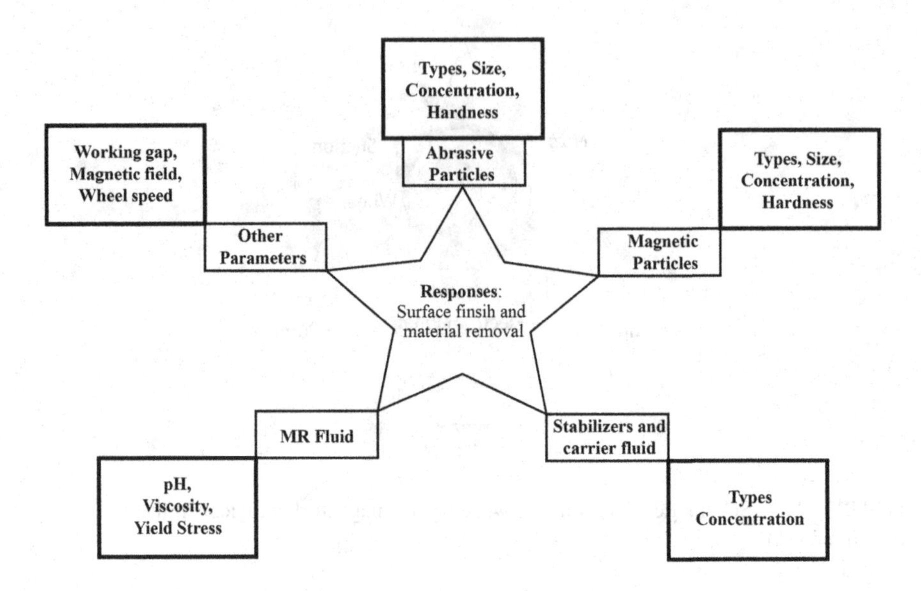

FIGURE 3.3 Process parameters of magnetorheological finishing.

The magnetic flux density required in the MRF process can be achieved by using either a permanent magnet or an electromagnet. Both permanent magnet and electromagnet have their own pros and cons regarding their use in the MRF process. While a permanent magnet setup is small, it cannot be used for in-process variation of magnetic flux density. In a permanent magnet MRF setup, the magnet must be changed for altering the magnetic flux density. In case of an electromagnet, although it is bulky, the magnetic flux density can be altered in-process by altering the magnitude of current supplied to the electromagnet windings.

The stiffness or the yield stress of the MRP fluid is a function of magnetic flux density. Several researchers have studied the effect of magnetic flux density on material removal rate (MRR) and yield stress of the MRP fluid. In one study, Ginder et al. [5] found the yield stress of the MRP fluid to be directly proportional to the square of the magnetic field strength. This is because a high magnetic field leads to an increase in the dipole moment between two CIPs, which causes them to stick or bond together strongly [6]. This results in an increase in the strength of the CIPs' chain links, hence, the yield stress of the MRP fluid increases causing the MRP fluid to stiffen. As a result of this, high finishing forces are encountered at higher magnetic flux density leading to the increase in the MRR of the workpiece. One of the adverse possibilities of finishing under high flux density is the chance of developing scratches on workpiece surfaces due to higher finishing forces.

3.2.1.2 Carbonyl Iron Particle Concentration

Carbonyl iron particles (CIPs) represent the magnetic phase of the MRP fluid and are responsible for imparting rheological changes to the MRP fluid under an external magnetic field. Researchers have found that with an increase in the concentration of CIPs in the MRP fluid, the yield stress and MRR increase while the final average surface roughness (Ra) decreases [7]. The CIPs attach themselves one after the other in the direction of the magnetic field lines when they are subjected or exposed to external magnetic field. Therefore, higher concentration of CIPs results in the formation of thicker/denser chains thereby increasing the stiffness or yield stress of the MRP fluid. Dense columnar chains of CIPs have the capability to resist higher forces required to break them, hence, higher yield stress is exhibited by the MRP fluid with increased concentration of CIPs [8].

During the formation of CIP chains or aggregation of CIPs in the presence of magnetic fields, abrasive particles are gripped between them. These gripped abrasives, when moved relative to the workpiece surface, smooth the roughness peaks and remove the material from the workpiece surface. Therefore, both final Ra and MRR depend on the strength at which these CIPs hold the abrasives. Higher CIP concentration aids in better gripping of abrasives. During the finishing operation, tightly held abrasives are prevented from rolling or sliding away. Hence, they perform better cutting action leading to an increase in the MRR and decrease in the final average surface roughness.

3.2.1.3 Abrasive Particle Concentration

Abrasive particles are responsible for imparting the finishing capability to the MRP fluid. The effect of abrasive particle concentration on yield stress of MRP fluid, MRR, and final Ra was studied by Sidpara and Jain [7]. They reported an optimum level of abrasive concentration for MRR and final Ra and a decrease in the yield stress of MRP fluid with an increase in the abrasive particle concentration. A higher number of abrasives hinders the CIP chain formation resulting in short and discontinuous CIP chains as opposed to long, continuous chains. Also, when an abrasive particle comes in between two magnetic particles (CIPs), the interaction force between the magnetic particles decreases, resulting in lower yield stress. Hence, high-abrasive concentration in the MRP fluid decreases the yield stress.

Shorey et al. [9] reported that abrasive particles are added in the MRP fluid to increase the MRR. However, Sidpara and Jain [7] found this to be true up to a certain concentration of abrasive particles only; after that higher abrasive concentration resulted in a decrease in the MRR. Initially, when the abrasive concentration is increased, they lead to an increase in MRR and low Ra. However, beyond a certain concentration, they weaken the CIP chains, resulting in a decrease of the yield stress. The MRP fluid, with low yield stress, fails to provide the necessary finishing action. As a result of this, the MRR decreases, and the final Ra increases.

3.2.1.4 Carrier Wheel Speed

In a wheel-type MRF system, the MRP fluid is poured on a rotating wheel with an electromagnet or permanent magnet below it. The speed of interaction of the magnetically energized MRP fluid with the workpiece surface is known as the carrier-wheel speed. Due to the magnetic flux density originating from the electromagnet or the permanent magnet placed below the carrier wheel, the MRP fluid sticks to the surface of the rotating wheel and moves with the same velocity as that of the rotating wheel (assuming no slip condition at interface of MRP fluid and carrier). Any change in the velocity of the carrier wheel changes the velocity of the MRP fluid too.

The MRR increases with the increase in the carrier-wheel speed. This is because, at higher velocity, the fluid interaction with the workpiece surface increases. The fast-moving abrasives apply greater shear force on the workpiece surface. As a result, enhanced shearing of the roughness peaks occur. Apart from this, high velocity also increases the frequency of the abrasives that participate in the finishing action. This also aids in enhancing in the MRR of the process.

3.3 BALL END MAGNETORHEOLOGICAL FINISHING

Ball end magnetorheological finishing (BEMRF) is one of the variants of MRF processing that is among the latest and the most advanced methods for finishing of three-dimensional (3D) and freeform surfaces [10]. As the name reflects, this process is also based on MRP fluid and utilizes the smart rheological effect of MRP fluid under the magnetic field to carry out the finishing action.

The highlight of this process is the BEMRF tool, which consists of a vertical spindle placed in the axial hollow cavity of a stationary electromagnet. The spindle is made up of magnetically soft or ferromagnetic material so that it gets magnetized easily when the electromagnet is energized. A hollow cavity is drilled throughout the spindle to allow the MRP fluid to be passed or pumped through it. The upper end of the spindle has a rotary valve that allows the entry of the MRP fluid. A peristaltic pump is used to push the MRP fluid to the tip of the spindle, the fluid passing through the rotary valve, and the centrally drilled hollow cavity of the spindle. As the MRP fluid reaches the spindle tip, the electromagnet is energized using a direct current (DC) power supply. When the electromagnet is energized, a magnetic field is produced. This magnetic field magnetizes the ferromagnetic spindle placed in the cavity of the electromagnet, and the spindle starts to behave like a permanent magnet. The magnetized spindle generates a hemispherical shape of magnetic flux lines at its tip, and because of this, the MRP fluid takes the shape of a hemispherical ball at the spindle's tip as shown in Figure 3.4.

At this energized state, the MRP fluid rotates and moves with the spindle. This hemispherical ball of MRP fluid containing polishing grade abrasives is moved relative to the workpiece surface to perform the finishing action. The stiffness of this hemispherical ball-shaped MRP fluid can be controlled by altering the intensity/magnitude of the current supplied to the electromagnet.

Also, the hemispherical ball of the MRP fluid is flexible in nature due to which it adjusts its compliance according to the workpiece profile. It is due to this property of compliance adjustment that makes this process capable of finishing 3D complex surfaces [11]. The movement of the hemispherical ball-shaped MRP fluid along the workpiece surface is facilitated by a computer numerical control (CNC) system [12]. The MRP fluid is replenished by switching off the current supplied to the electromagnet and pumping the fresh fluid through the hole drilled in the tool spindle.

3.3.1 Mechanism of Material Removal in BEMRF Process

The homogeneity of the MRP fluid is lost during the finishing operation. At the tip of the spindle, a rich composition of CIPs is observed, whereas at the workpiece surface the concentration of abrasives are high. This is because magnetically opposite

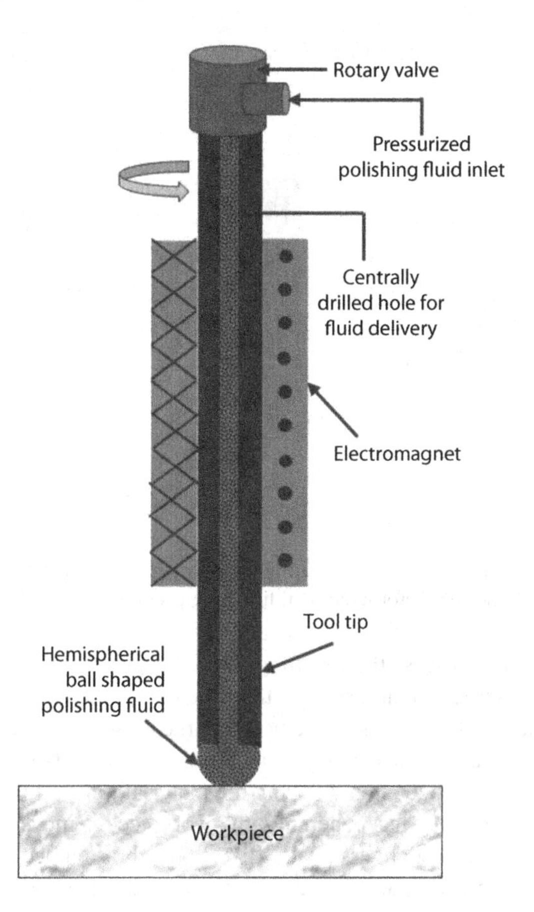

FIGURE 3.4 Schematic of ball end magnetorheological finishing process.

poles of different strengths are formed in the BEMRF process; one is at spindle's tip because of the electromagnet, and the other is at the workpiece surface, which is magnetically induced by the spindle (Figure 3.5). The magnetic pole at the spindle's tip is nearer to the electromagnet, and therefore it is stronger than that formed at the workpiece's surface. This difference in the strength of the two poles results in the formation of magnetic flux density gradient in the working gap between the tool tip and the workpiece surface [13]. The constituents of the MRP fluid present in this working gap react differently to this magnetic gradient. The ferromagnetic particles, being magnetic in nature, rush towards the spindle's tip and push the non-magnetic abrasive particles towards the lower flux density region, i.e., workpiece surface. This action of abrasives being pressed towards the workpiece surface forms the basis of normal force acting on the workpiece during the finishing operation.

The stiffness or the yield stress of the MRP fluid present in the working gap determines the gripping strength of the abrasives between the CIP chains. In the magnetically energized state, the columnar chains of CIP possess the capability to resist forces required to break them. This gives the abrasive particles entrapped between the CIP chains the required shear force to remove/cut the roughness peaks.

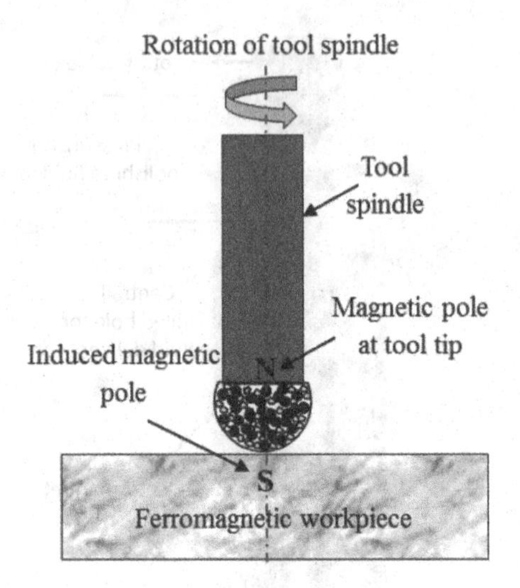

FIGURE 3.5 Magnetorheological polishing fluid magnetization between two magnetic poles formed in ball end magnetorheological finishing process.

Hence, in BEMRF process the finishing force comprises normal and shear force. Due to the normal force, the abrasive particles penetrate the workpiece surface, and due to the shear force, these abrasives remove/cut the roughness peaks on the surface. The material removal takes place in the form of tiny chips. Both two-body abrasion and three-body abrasion are encountered during the material removal in BEMRF process [14]. When the abrasive that has penetrated the workpiece is moved, it is opposed by a resistive force from the roughness peaks. Two-body abrasion occurs when the shear force applied by the abrasive particle is greater than the resistive force, and the abrasive particle cuts the roughness peaks and moves ahead as seen in Figure 3.6(a). However, at times the resistive force is higher than the shear force. This may be because of the low yield stress of the MRP fluid, or because the abrasive particle has indented too deeply into the workpiece surface. In this case, the CIP chains break and the abrasive is rolled over the roughness peaks, causing three-body abrasion (Figure 3.6b).

3.3.2 BEMRF Tool

The BEMRF tool is an essential component that gives the BEMRF process its unique identity among the various types of MRF processes. The tool is an assembly of a variety of components like electromagnet, spindle, tool tip, servomotor, rotary valve, etc. [15]. Figure 3.7 shows the schematic diagram of the BEMRF tool assembly.

The major components of the tool are as follows:

a. *Spindle*: The cylindrical spindle forms the rotating part of the tool and is placed vertically in the axial hollow cavity of the electromagnet. The spindle is made up of magnetically soft or ferromagnetic material so that it gets magnetized easily when the electromagnet is energized. A hollow cavity is drilled throughout the spindle to allow the MRP fluid to be passed

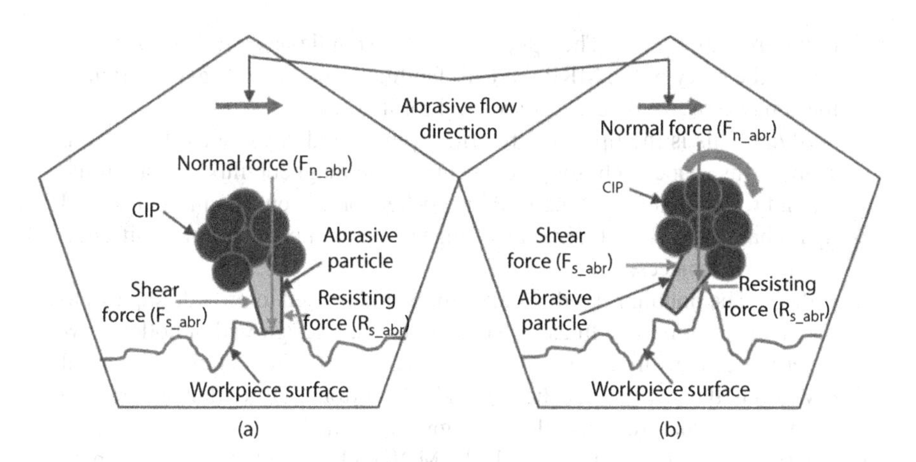

FIGURE 3.6 Abrasive wear mechanism in ball end magnetorheological finishing process: (a) two-body and (b) three-body.

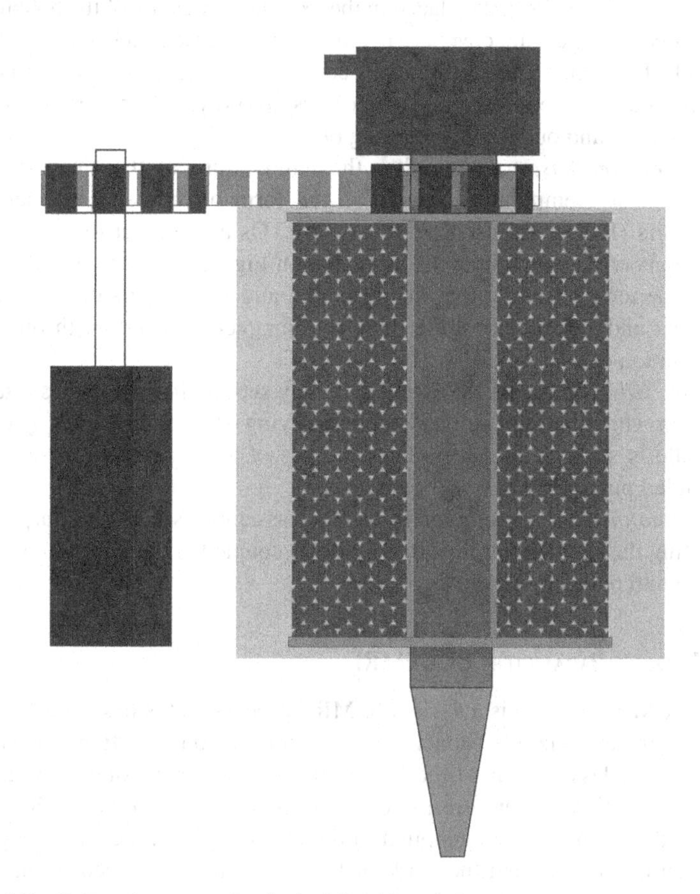

FIGURE 3.7 Ball end magnetorheological finishing tool.

or pumped through it. The upper end of the spindle has a rotary valve that allows the entry of the MRP fluid. To facilitate smooth and easy rotation of the spindle, two ball bearings are used for its support.

b. *Tool tip*: This is the tip of the rotating spindle and is a conical frustum in shape. This shape is chosen to concentrate the magnetic flux density at the tip and to facilitate finishing of deep pockets or narrow channels. The tool tip is changeable, and its size can change to accommodate or finish different sizes of workpieces.
c. *Bobbin*: The bobbin is made of aluminum and is used to provide support to the electromagnet windings. It has a cylindrical design with a hollow cavity at the center through which the spindle passes. The hollow cylindrical cavity has flanges both at the top and the bottom. These flanges act as a supporting structure to the electromagnet windings.
d. *Electromagnet*: The highlight of the BEMRF tool assembly is the electromagnet that consists of multi-turn and multi-layers of copper wire wound over the bobbin. The length of the electromagnet is the same as that of the bobbin. When direct current passes through the copper coils, the electromagnet magnetizes the vertical spindle placed in the axial hollow cavity of the bobbin.
e. *Outer cover*: The outer cover is a nylon-made cylindrical jacket that encloses the electromagnet and facilitates its cooling by keeping the electromagnet immersed in the cooling oil. Two nozzles are provided on the outer cover for the inlet and outlet of the cooling oil.
f. *Temperature sensor*: To monitor the temperature of the electromagnet continuously, temperature sensors in the form of resistance temperature detectors (RTDs) are used. Pt-100-type RTDs are used at three different locations of the electromagnet as shown on Figure 3.7. These sensors provide the temperature at the inner, middle, and outer layers of the electromagnet and aid in controlling the electromagnet temperature through the circulation of cooling oil.
g. *Rotary valve*: In the BEMRF tool assembly set-up, the spindle has rotation motion while the fluid delivery tube supplying MRP fluid to the top of the spindle is static. This necessitates the use of a rotary valve to serve the intended purpose.
h. *Servomotor*: The servomotor in the tool assembly serves the purpose of rotating the spindle. For this, the spindle is coupled to the servomotor using a belt and pulley mechanism.

3.3.3 Process Parameters of BEMRF

Since the BEMRF process is a variant of MRF process and is based on MRP fluid, the fluid parameters viz. CIP concentration, abrasive concentration, abrasive size, etc. are more or less similar to the MRF process and were extensively discussed in the earlier sections. There are three process parameters on which the BEMRF depends, and they are current supplied to the electromagnet, the working gap, i.e., gap between the tool tip and the work surface and rotational speed of the spindle. All these factors, along with the fluid parameters, play a crucial role in getting the optimum performance of the BEMRF process.

3.3.3.1 Electromagnet Current

In BEMRF process, the current supplied to the electromagnet to magnetize the rotating spindle is known as the electromagnet current. The tip of the magnetized spindle acts as a magnetic pole and is responsible for producing magnetic flux density in the gap below it. The magnetic flux density is directly proportional to the magnitude of current supplied to the electromagnet.

Researchers have studied the influence of electromagnet current on the percentage reduction in surface roughness (%Δ Ra) and the finishing forces associated with BEMRF process. Percentage reduction in surface roughness (%Δ Ra) is calculated by finding the difference between the initial and the final average surface roughness and dividing it by the initial average surface roughness. With an increase in the electromagnet current, %Δ Ra increases as a higher cutting action is observed. This is because at higher current, high magnetic flux density causes better gripping of abrasives by the CIP chains. These tightly held abrasives possess superior cutting capability and hence %Δ Ra increases.

The finishing forces in BEMRF process are the normal and shear force. Alam et al. [16] found these forces increase with increase in the electromagnet current, and they reported that this behavior was due to these forces having a direct relationship with the magnetic flux density. The increase in the electromagnet current supplied to the BEMRF tool increases the magnetic flux density, and hence the forces increase.

3.3.3.2 Working Gap

The space below the frustum-shaped tool tip and above the workpiece surface is filled with energized MRP fluid on the BEMRF process. This space is commonly referred to as the working gap. The tool tip comes closer to the workpiece surface if the working gap is reduced and vice versa. On either side of the working gap, two opposite magnetic poles are formed. When the working gap is reduced, the two opposite poles come closer to each other, thereby enhancing the magnetic flux density in the working gap. The flux density weakens when the poles are moved further apart because of an increased working gap.

Since the working gap alters the magnetic flux density, both %Δ Ra and the finishing forces increase with the reduction in the working gap and vice versa. The reason for this is the same as discussed in the Section 3.3.3.1.

3.3.3.3 Spindle Speed

In BEMRF process, the MRP fluid sticks to the tip of the spindle because a strong magnetic pole is formed at the tool tip due to the magnetization of the spindle. The MRP fluid rotates with the same speed as that of the spindle as there is no slip between them. As the rotational speed of the spindle is increased, the centrifugal force acts on the solid constituents of the MRP fluid. The magnitude of centrifugal force is higher at the peripheral region of the circular tool tip. While the magnetic CIP particles are held together by the magnetic field, the non-magnetic abrasive particles get dragged away from the iron particle chains. This leads to a reduction in %Δ Ra as the rotational speed of the spindle increases. Also, the centrifugal force is proportional to the square of the rotational velocity, so a little change in rotational speed imparts a big change in centrifugal force.

The effect of spindle speed on the finishing forces was studied by Alam et al. [16]. They found the normal and shear force react differently with the change in the spindle speed. As the rotational speed of the spindle increases, the speed of the gripped abrasive also increases. Therefore, the fast-moving abrasive applies greater force to shear off the roughness peaks and hence the shear force increases. However, an increase in rotational speed leads to a decrease in value of normal force. At higher rotational speeds, the abrasive and the CIP chains experience greater centrifugal force. While this centrifugal force aids in the increment of shear force, it proves detrimental in the case of normal force. In BEMRF process, the magnetic flux density in the working gap also decreases radially from the center to the periphery of the tool tip [14]. When the rotational speed of the tool increases, the MRP fluid towards the periphery of the tool starts yielding. This leads to the destruction of the CIP chains below the tool's peripheral region and hence the normal force decreases due to breaking of the CIP chains, which are not able to transfer the force to the non-gripped/loose abrasive particles.

3.3.4 Mathematical Modeling of BEMRF Process

Singh et al. [13] proposed a model for normal force considering the equivalent magnetic circuit of the BEMRF setup. They also explained the material removal mechanism based on the normal force. By assuming the total volume of MRP fluid in the working gap to be V, they calculated the number of CIPs in the engaged MRP fluid volume, which is given by:

$$N_{CIP} = \frac{\%\ volume\ fraction\ of\ CIP\ in\ MRP\ fluid \times V}{volume\ of\ single\ CIP\ particle} \tag{3.1}$$

The total number of CIP chains is calculated by dividing the number of CIPs in the MRP fluid volume by the number of CIPs required to form a single chain. The abrasives that participate in cutting are known as active abrasives. The number of active abrasives on the workpiece surface [13] is given by:

$$N_{AAB} = \frac{\%\ volume\ fraction\ of\ abrasive\ in\ MRP\ fluid \times A_Z \times diameter\ of\ abrasive\ particle}{volume\ ofsingle\ abrasive\ particle} \tag{3.2}$$

where A_Z is the projected area of the tool tip. After finding the number of CIPs and number of active abrasives in the MRP fluid volume, Singh et al. [13] calculated the total flux flow considering the equivalent magnetic circuit of the BEMRF process as shown in Figure 3.8. The total flux flow is given by:

$$\emptyset = \frac{NI}{\left(R_1 + R_2 + R_Z + R_W + \frac{R_{air}}{2}\right)} \tag{3.3}$$

where N and I are the number of turns and current in the electromagnet, respectively, and R is the reluctance of the various components of the magnetic equivalent circuit as shown in Figure 3.8.

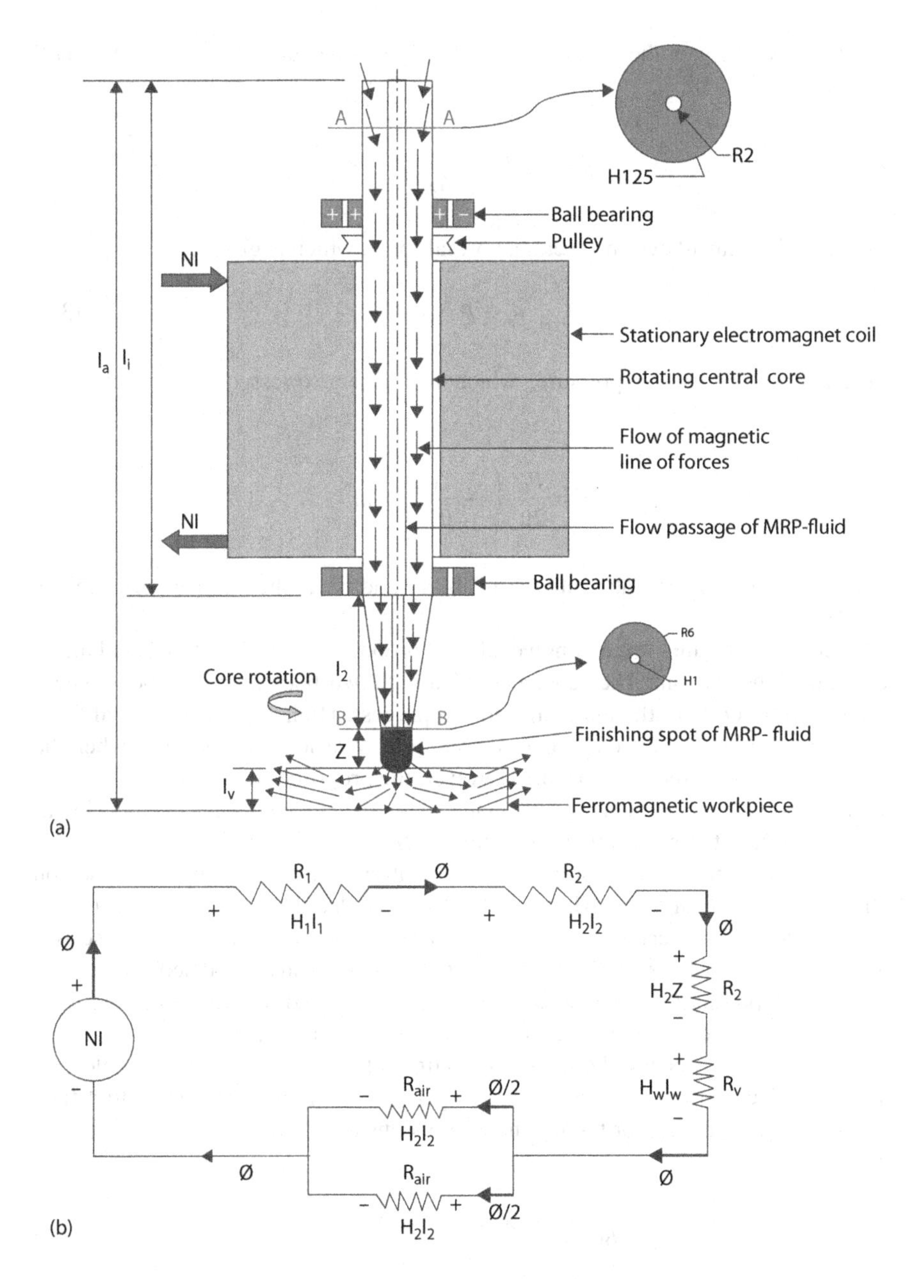

FIGURE 3.8 Schematic of (a) flow of magnetic flux and (b) equivalent magnetic circuit for flow of magnetic flux in ball end magnetorheological finishing process (From ref. [13].)

In the working gap, the magnetic flux density was calculated by Singh et al. [13] by:

$$B_Z = \frac{\emptyset}{A_Z} \tag{3.4}$$

From this, they calculated the average normal force, which is given by:

$$F_n = P \times A_Z \tag{3.5}$$

where P is the polishing pressure, which is further expressed by the following equation:

$$P = \frac{B_Z{}^2}{2\mu_o}\left(1 - \frac{1}{\mu_{rz}}\right) N/m^2 \tag{3.6}$$

where μ_o is the free air's absolute permeability and μ_{rz} is the relative permeability of MRP fluid.

Based on the normal force, Singh et al. [13] proposed the surface finish and material removal mechanism. They concluded that both two-body and three-body mechanisms played a role in the material removal process. Higher value of normal force resulted in two-body abrasion, whereas three-body abrasion was dominant when the normal force was of the lower value. However, the proposed wear theory (two-body and three-body abrasive wear) and the material removal mechanism lacked the inclusion and the effect of shear force associated with the BEMRF process.

To have an improved understanding of the material removal mechanism based on both normal and shear force, Alam and Jha [14] modeled the normal and shear force and proposed the material removal mechanism considering the role of these two forces simultaneously (explained in Section 3.3.1). They also modeled the surface roughness to predict the final surface roughness value and the finishing time for a mild steel workpiece involving certain machining parameters.

In the BEMRF process, the forces are mainly affected by current in the electromagnet producing the magnetic flux density in the working gap. Hence, first they [14] established a model for the magnetic flux density in the working gap, which is given by:

$$\vec{B}(z) = \frac{\mu_0 \times m_{tool}}{4\pi z^2} + \frac{\mu_0 \times m_{wp}}{4\pi (t - z)^2} \tag{3.7}$$

where m_{tool} and m_{wp} are the pole strength of the magnetic poles formed at the tool tip and the workpiece surface, t is the working gap, and z is the axial distance from the tool tip. This variation of flux density in the working gap includes the effect of induced magnetization of workpiece by the magnetic pole formed at the tool tip. Therefore, it predicts a more realistic picture of magnetic flux density variation in the working gap.

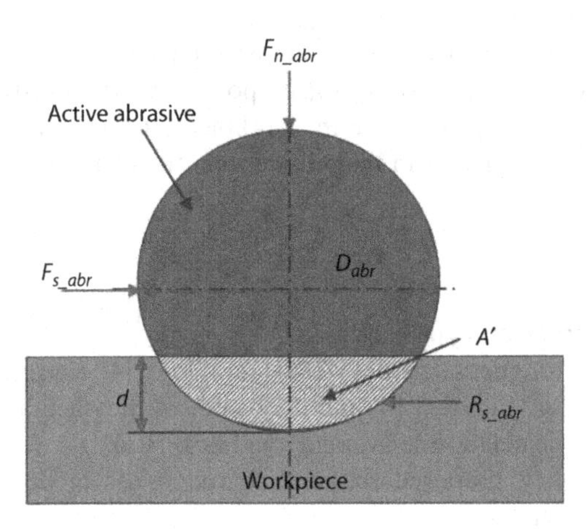

FIGURE 3.9 Finishing forces on active abrasive particle in ball end magnetorheological finishing process.

Assuming the CIPs and abrasives form a half BCC (body-centered cubic) structure and that for a single active abrasive on the workpiece surface, there are four magnetic CIPs surrounding it, Alam and Jha [14] found the normal force acting on the active abrasive particle to be a function of magnetic flux density, which is given by:

$$F_{n\_abr}(z) = \frac{m\chi_{m\_CIP}}{\mu_o} B(z) \frac{dB(z)}{dz} \tag{3.8}$$

where m and χ_{m_CIP} are CIP's mass and magnetic susceptibility, respectively. The normal force in BEMRF process depends not only on the magnitude of the flux density but also on the gradient of the flux density in the working gap. The presence of ferromagnetic material as the workpiece enhances the magnetic flux density in the working gap, and therefore a higher value of normal force is realized while finishing magnetic materials as compared to non-magnetic materials.

The shear force acting on the active abrasive particle in BEMRF process as proposed by Alam and Jha [14] is given by:

$$F_{s\_abr} = (A - A')\tau_y \tag{3.9}$$

where A and A' are the projected area of the active abrasive particle and the indented portion of the active abrasive in the workpiece surface as shown in Figure 3.9, and τ_y is MRP fluid's yield stress.

The projected area A' of the indented abrasive particle [17] is expressed as:

$$A' = \frac{D_{abr}^{\ 2}}{4} \sin^{-1} \frac{2\sqrt{d(D_{abr} - d)}}{D_{abr}} - \sqrt{d(D_{abr} - d)}\left(\frac{D_{abr}}{2} - d\right) \tag{3.10}$$

where D_{abr} is the diameter of the active abrasive particle, and d is the indentation depth of the abrasive particle. They [14] proposed that the finishing action or the shearing of the roughness peaks from the workpiece surface only takes place when the shear force F_{s_abr} is greater in magnitude than the resistive force R_{s_abr}, which is given by:

$$R_{s\_abr} = A' \times \sigma_y \tag{3.11}$$

where σ_y is the workpiece's yield strength.

For the surface roughness model, Alam and Jha [14] assumed the roughness peaks to be triangular (Figure 3.10) and the peak-to-valley height of the roughness peaks (Rt) is twice the center-line average roughness (Ra).

Based on this, they proposed the surface roughness model for the BEMRF process, which is given by

$$R_a^i = R_a^o - \frac{n_s \times A'}{l_w} \tag{3.12}$$

where R_a^o is the initial average surface roughness value, R_a^i is the final average surface roughness after i^{th} rotation, l_w is the distance travelled by the active abrasive during a single rotation of the tool, which is equal to $2\pi R_m$ (Figure 3.11 schematically

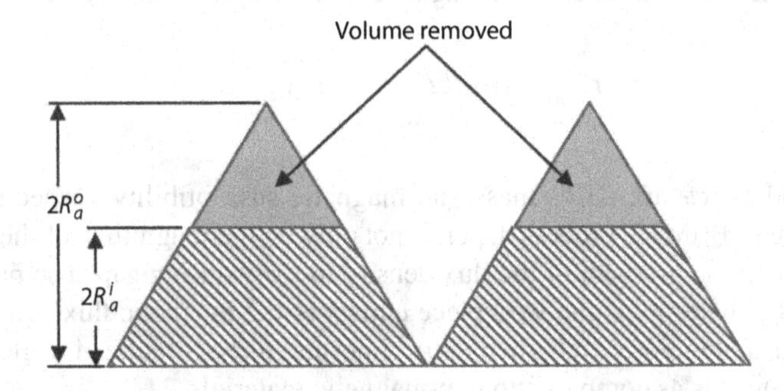

FIGURE 3.10 Schematic of the assumed triangular surface roughness peaks.

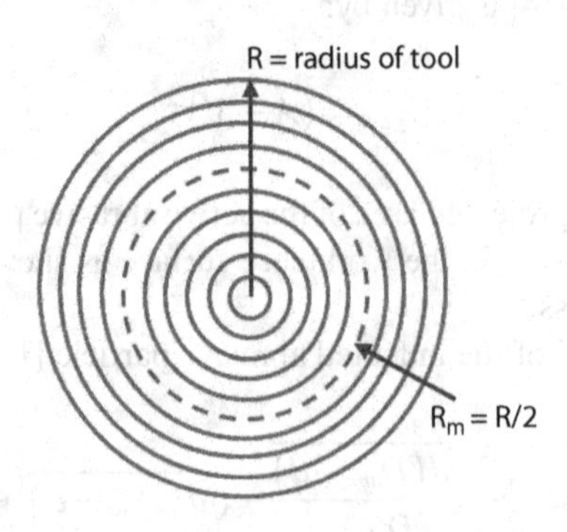

FIGURE 3.11 Distance travelled by the active abrasive during single rotation of tool.

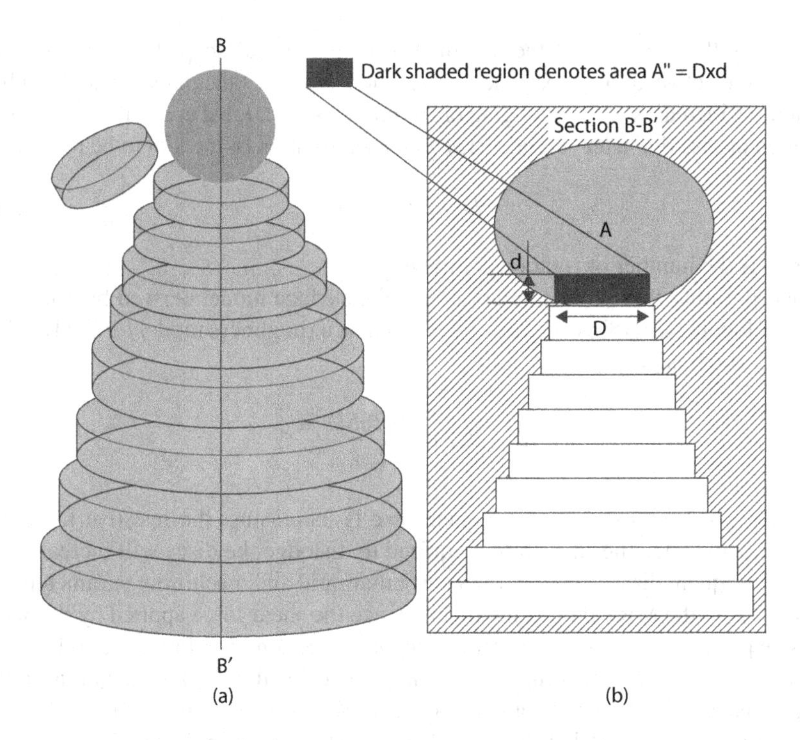

FIGURE 3.12 Conical shape of roughness peak dissected into multiple discs where the first disc is disc is dislodged by the active abrasive particle and (b) cross sectional view along B-B′ showing area A″ dislodged by active abrasive particle.

shows R_m), and n_s is the number of active abrasives indenting the work surface per rotation of the BEMRF tool.

The mathematical model to calculate the surface roughness of the BEMRF process was validated experimentally by Alam and Jha [14] at 2, 3, and 4 amperes of electromagnet current. The error between the experimental and theoretical values is found lying within 7.23% (for 2A current) to 31.19% (for 4A current). They concluded that for the same rotations of tool spindle, an increase in the magnetizing current leads to a decrease in the final surface roughness value. Therefore, the maximum reduction in roughness value for mild steel workpiece is obtained at 4A current. At this current value, the average roughness of mild steel sample reduced from an initial value of 286 nm to final value of 218 nm for 2000 rotations of tool spindle.

Iqbal et al. [18] modeled the transient behavior of surface roughness reduction in BEMRF process. For this they assumed the roughness peaks to be in the shape of a cone that is dissected into several discs as shown in Figure 3.12(a).

The part (b) of Figure 3.12 shows the area dislodged by the active abrasive particle which is mathematically represented as:

$$A'' = D \times d \tag{3.13}$$

where D is the diameter of the assumed discs of conical shaped roughness peaks and d is the active abrasive particle's indentation depth. As the diameter of the subsequent discs increases with dislodging of the previous disc, Iqbal et al. [18] assumed the variation of disc diameter (Figure 3.13) with each rotation of the tool to be related as:

$$D = 2.5 \times i \times d \tag{3.14}$$

where i is the number of rotations of the tool.

Incorporating these two assumptions in the surface model derived by Alam and Jha [14], Iqbal et al. [18] proposed a revised surface roughness model for the BEMRF process, which is given by:

$$R_a^i = R_a^o\left(1 - \frac{0.785D^2 id}{l_w n_s A''}\right) \tag{3.15}$$

Since the diameter of every subsequent disc is increasing, the resistive force will increase, and hence the shear force required to dislodge the discs will increase with every subsequent disc. However, for a particular fluid and machining parameters, the yield stress of the MRP fluid is fixed, and hence the shear force applied by the active abrasive particle is constant. Therefore, beyond a certain disc diameter, no finishing action takes place, and this disc is known as the critical disc. The height up to this critical disc is known as the critical roughness as shown in Figure 3.14.

Alam et al. [19] modeled the size of the finishing spot in BEMRF process. For a certain area of the workpiece requiring localized/selective finishing, the mathematical model predicts the size of the required finishing spot. The energized MRP fluid forms a hemispherical ball at the tip of the tool and for a particular value of electromagnet current and spindle speed, the size of the finishing spot depends upon the axial distance at which workpiece makes a contact with the hemispherical ball-shaped fluid. Therefore, a larger spot size (r_1) is obtained if the workpiece surface is

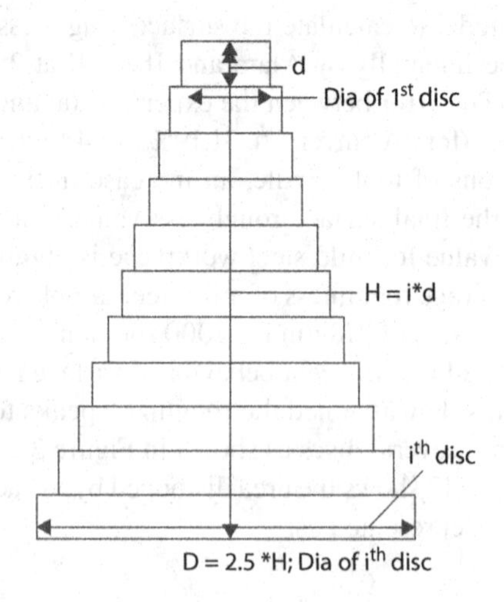

FIGURE 3.13 Variation of disc diameter with each rotation of the tool.

FIGURE 3.14 Schematic of the critical disc and the critical roughness.

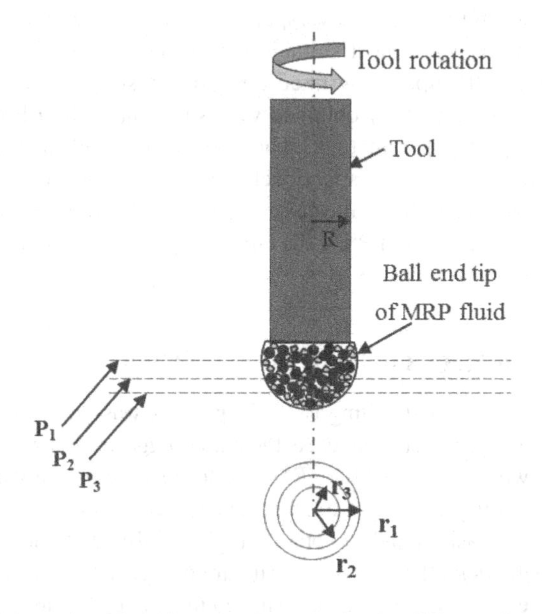

FIGURE 3.15 Schematic of finishing spot sizes obtained at different distances from tool tip.

near to the tool at plane P_1 and a smaller spot size (r_2) if the workpiece surface is far away from the tool at plane P_3 as shown in Figure 3.15.

Assuming the energized MRP fluid is Bingham fluid, and balancing the centrifugal force of the rotating fluid with the yield stress of the MRP fluid, Alam et al. [19] modeled the size of the finishing spot, which is given by:

$$r = \sqrt{\frac{2 \times \tau_y}{\rho_{MR} \times \omega^2}} \tag{3.16}$$

where r is the radius of the finishing spot size, ρ_{MR} is MRP fluid's density, and ω is the angular velocity of the rotating MRP fluid. The density of MRP fluid is calculated as:

$$\rho_{MR} = \begin{pmatrix} \rho_{CIP} \times volume\ percentage\ of\ CIP + \\ \rho_{abr} \times volume\ percentage\ of\ abrasive + \\ \rho_{med} \times volume\ percentage\ of\ medium \end{pmatrix} /100 \quad (3.17)$$

where ρ_{CIP} is the density of CIP, ρ_{abr} is the density of abrasive particle, and ρ_{med} is the density of the carrier medium of the MRP fluid.

3.3.5 Closed-Loop Control of BEMRF Process

A very small percentage of BEMRF archival literature focuses on automation; work reported by Iqbal and Jha [20] is one such case wherein they first carried out a time-based roughness reduction on an EN-31 sample using optimum machining parameter sets from their work in [21]. They developed a parameter selection algorithm (PSA) enabling the selection of the most suitable parameter set for effecting the highest percentage of ΔRa in the upcoming fixed-duration finishing cycle. The automation lies in using these experimentally obtained values to control the BEMRF process by dedicating an NC part program to it. The system involved an in-situ roughness measurement system for roughness feedback at the end of each cycle; the value thus obtained became the initial Ra for the next finishing cycle. Making the combined use of a dedicated part program and PSA, the BEMRF process is controlled in a closed-loop manner as shown in Figure 3.16.

3.4 MAGNETORHEOLOGICAL JET FINISHING

The magnetorheological jet finishing (MRJF) process was developed by Kordonski [22] in the early part of the last decade. As the name suggests, this process is one of the variants of MRF, which uses a jet of MRP fluid to impinge on the workpiece surface and provide the finishing action. An axial magnetic field is used to stabilize the MRP fluid jet containing abrasives and to obtain a highly collimated and coherent jet [23]. Beyond the magnetic field, the stabilized structure of the MRP fluid jet starts decreasing and the jet flares out. When the round-shaped magnetically stabilized jet impinges on the workpiece surface, the radial spread of the jet provides the required energy to the abrasive to shear off the roughness peaks [24]. The impingement of the MRP fluid on the workpiece surface leads to the generation of surface shear stress that is more than sufficient to cause removal of tiny bits of materials from the workpiece surface.

Losing its cohesiveness is a fundamental property of the fluid jet when it moves out of a nozzle. Hence, application of an axial magnetic field is employed in this process to stabilize the MRP fluid jet flowing out of the nozzle. While a jet of water, being less viscous than MRP fluid, remains stabilized for about 2 nozzle diameters, the MRP fluid jet remains coherent for about 7–8 diameters of nozzle in the absence of magnetic field. In the presence of an axial magnetic field, the MRP fluid

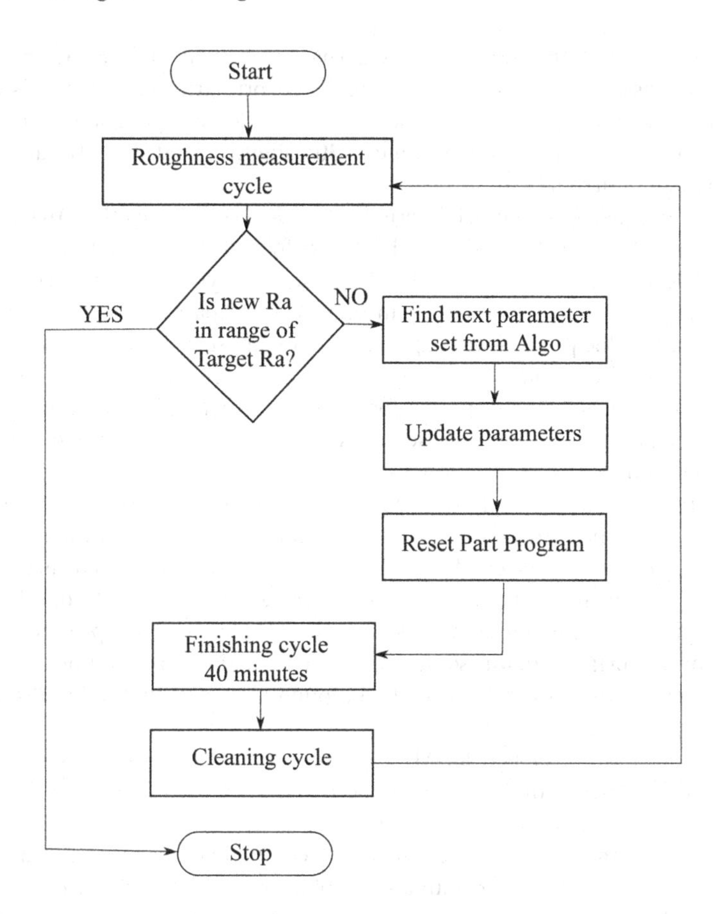

FIGURE 3.16 Flowchart for closed loop control of ball end magnetorheological finishing process.

jet remains coherent for more than 200 nozzle diameters before breaking down and spreading in the form of droplets.

Since its inception, MRJF process is primarily used for finishing of optical workpieces. In MRJF process, the finishing spot is formed at certain distance (some tens of centimeters) from the nozzle. Utilizing this property of the process, Kordonski et al. [25] finished steep concave surfaces and cavities of glass. In another study carried out by Tricard et al. [26], MRJF process was used for finishing of freeform, conformal, and steep concave optics that are otherwise difficult to finish by traditional finishing processes. They concluded that this process could also be used for finishing of single crystals, metals, and advanced ceramics.

3.5 APPLICATIONS

MRF and its allied processes have been employed to finish almost all types of materials ranging from glasses and ceramics to metals like steel, copper, aluminum, etc. It has also shown its capability to finish all types of surfaces like planar, non-planar, freeform, and internal and external cylindrical surfaces. Compared to its allied processes, MRF in particular has been primarily used for finishing of non-metallic materials especially glasses [9, 27–29]. Some applications of the MRF process includes finishing of silicon wafers [30], single crystal ZnS [31], and aluminum lightweight mirrors [32].

Researchers have tried to finish internal cylindrical surfaces by the MRF process. Bedi and Singh [33] developed the MRF process for finishing of traditionally honed cylindrical ferromagnetic components like hydraulic cylinders, cylindrical molds and dies, injection barrels of molding machines, etc. Apart from reducing surface roughness values, the process was capable of removing surface defects like torn and folded metals, cavities, holes, honing grooves, etc. Grover and Singh [34] developed a magnetorheological honing process for finishing of variable cylindrical internal surfaces. The permanent magnet-based finishing tool could be adjusted radially as per the internal diameter of the cylindrical workpiece.

MRF processes are not only used for finishing of internal cylindrical surfaces (magnetorheological honing), but their applications are also seen in nanofinishing of external cylindrical surfaces like cylinders, grooves, tapers, stepped surfaces, etc. Sadiq and Shunmugam [35] combined honing and MRF processes to develop magnetorheological abrasive honing processes (MRAH) and used this process to finish external curved surfaces. In this setup, reciprocating and rotation motion is imparted to the workpiece where the role of rotation motion to obtain surface finish is higher compared to reciprocating motion.

Singh et al. [36] developed an MRF process for external cylindrical surfaces. which mimicked the turning process in operation. They used a modified MRF tool with a flat and curved surface for finishing cylindrical workpieces. For 90 minutes of finishing time, the percent reduction in surface roughness values Ra, Rq, and Rz were 54.41%, 51.65%, and 40% with a flat tool tip and 80.88%, 81.32%, and 82.5% with a curved tool tip.

Using BEMRF process, Singh et al. [11] finished a typical 3D ferromagnetic workpiece surface made by a milling process at different angles of projection such as flat, 30°, 45°, and curved surfaces. They were able to reduce the surface roughness to as low as 16.6 nm, 30.4 nm, 71 nm, and 123.7 nm, respectively, on flat, 30°, 45°, and curved surfaces for 60 passes of finishing. However, the variation in surface finish can be minimized by providing a tilting motion to BEMRF finishing tool. This can be done by the addition of a rotary axis in the finishing setup, so that the tool tip surface can always be perpendicular to the 3D workpiece surfaces during the finishing operation. This will produce a uniform magnetic flux density zone irrespective of 3D workpiece surfaces and will result in uniform finishing over 3D surfaces.

Alam et al. [37] mounted the BEMRF tool assembly on a five-axis CNC setup to enhance the finishing performance for 3D surfaces. The five-axis configuration facilitates the ball end fluid tip to conform to the surface profile of the workpiece and hence provide a better finished surface. Using this five-axis CNC BEMRF machine with a customized controller, they achieved a uniform finish with a higher percent reduction in Ra value on a freeform mild steel component.

Several researchers have successfully attempted to finish a variety of non-magnetic materials using BEMRF process. Saraswathamma et al. [38] used deionized water as a carrier medium and cerium oxide as an abrasive to finish the silicon wafer using the BEMRF process. Singh et al. [39] polished fused silica glass using cerium oxide Cerox 1663 abrasive powder. In another application of BEMRF process, Khan et al. [40] finished polycarbonate using diamond abrasive particles and were able to achieve a final average surface roughness value of 30.5 nm with 54% improvement in surface roughness.

Khan and Jha [41] used BEMRF process to finish oxygen-free high-conducting (OFHC) copper and reduced the surface roughness of a flat workpiece from Ra = 65.90 nm to Ra = 38 nm. They placed a permanent magnet below the copper workpiece to enhance the magnet field strength and facilitate the finishing process. However, the use of a permanent magnet below the copper workpiece limits the process for medium and thin workpieces only. For a thick workpiece the permanent magnet fails to enhance the magnetic field strength at the top surface.

Khan and Jha [42] suggested the use of electrolytic iron powder (EIP) instead of CIPs for finishing of copper as larger-sized EIP exert greater forces on diamagnetic copper surface as compared to smaller-sized CIP. Incorporating the use of EIP instead of CIP in MRP fluid, Alam et al. [43] finished non-planar copper mirrors using a newly developed BEMRF tool with solid rotating core and slotted tip. The solid spindle core and multiple slots at the tip improve the magnetic flux density and grip the MRP fluid firmly during the finishing operation. This is essential for finishing of diamagnetic copper workpieces, which repels the magnetic field and weakens the MRP fluid brush above its surface. Two non-planar copper workpieces (convex and concave samples) were finished using the new tool mounted on a five-axis CNC setup to produce mirror-like reflective surfaces. In two finishing cycles with varying magnetizing current, the surface roughness of two samples were reduced from initial values of 120.1 and 129.2 nm to final values of 40.3 and 43.2 nm for convex and concave samples, respectively.

Khan et al. [44] finished aluminum using BEMRF process and reduced the average surface roughness of the workpiece from 91 nm to 42.2 nm in 32 minutes (8 finishing loops) of finishing time. The topographical analysis of the finished aluminum surface showed that the light scratches are completely removed from the surface and the deep scratches turned in light discontinuous scratches after 32 minutes of finishing. The finished surface was free from embedment of abrasive particles.

REFERENCES

1. Jain, V. K. (Ed.). (2016). Nanofinishing Science and Technology: Basic and Advanced Finishing and Polishing Processes. CRC Press.
2. Sidpara, A., Das, M., & Jain, V. K. (2009). Rheological characterization of magnetorheological finishing fluid. Materials and Manufacturing Processes, 24(12), 1467–1478.
3. Kordonski, W. I., & Jacobs, S. D. (1996). Magnetorheological finishing. International Journal of Modern Physics B, 10(23n24), 2837–2848.
4. Jain, V. K. (2009). Magnetic field assisted abrasive based micro-/nano-finishing. Journal of Materials Processing Technology, 209(20), 6022–6038.
5. Ginder, J. M., Davis, L. C., & Elie, L. D. (1995). Rheology of magneto-rheological fluids: Models and measurements. In 5th International Conference on ER Fluids and MR Suspensions (pp. 504–514). Singapore: World Scientific.

6. Huang, J., Zhang, J. Q., & Liu, J. N. (2005). Effect of magnetic field on properties of MR fluids. International Journal of Modern Physics B, 19(01n03), 597–601.
7. Sidpara, A., & Jain, V. K. (2014). Rheological properties and their correlation with surface finish quality in MR fluid-based finishing process. Machining Science and Technology, 18(3), 367–385.
8. Alam, Z., Khan, D. A., Iqbal, F., & Jha, S. (2019). Effect of polishing fluid composition on forces in ball end magnetorheological finishing process. International Journal of Precision Technology, 8(2–4), 365–378.
9. Shorey, A. B., Jacobs, S. D., Kordonski, W. I., & Gans, R. F. (2001). Experiments and observations regarding the mechanisms of glass removal in magnetorheological finishing. Applied Optics, 40(1), 20–33.
10. Singh, A. K., Jha, S., & Pandey, P. M. (2011). Design and development of nanofinishing process for 3D surfaces using ball end MR finishing tool. International Journal of Machine Tools and Manufacture, 51(2), 142–151.
11. Singh, A. K., Jha, S., & Pandey, P. M. (2012). Nanofinishing of a typical 3D ferromagnetic workpiece using ball end magnetorheological finishing process. International Journal of Machine Tools and Manufacture, 63, 21–31.
12. Alam, Z., Iqbal, F., & Jha, S. (2015). Automated control of three axis CNC ball end magneto-rheological finishing machine using PLC. International Journal of Automation and Control, 9(3), 201–210.
13. Singh, A. K., Jha, S., & Pandey, P. M. (2013). Mechanism of material removal in ball end magnetorheological finishing process. Wear, 302(1–2), 1180–1191.
14. Alam, Z., & Jha, S. (2017). Modeling of surface roughness in ball end magnetorheological finishing (BEMRF) process. Wear, 374, 54–62.
15. Khan, D. A., Alam, Z., Iqbal, F., & Jha, S. (2020). Design and Development of Improved Ball End Magnetorheological Finishing Tool with Efficacious Cooling System. In Advances in Simulation, Product Design and Development (pp. 557–569). Springer, Singapore.
16. Alam, Z., Khan, D. A., Iqbal, F., & Jha, S. (2017). Analysis of forces in ball end magnetorheological finishing process. In Proceedings of the 39th MATADOR Conference. U.K.: University of Manchester.
17. Jain, R. K., Jain, V. K., & Dixit, P. M. (1999). Modeling of material removal and surface roughness in abrasive flow machining process. International Journal of Machine Tools and Manufacture, 39(12), 1903–1923.
18. Iqbal, F., Alam, Z., & Jha, S. (2020). Modelling of transient behaviour of roughness reduction in ball end magnetorheological finishing process. International Journal of Abrasive Technology, 10(3), 170–192.
19. Pramanik, A. (Ed.). (2021). Machining and Tribology: Processes, Surfaces, Coolants, and Modeling. Elsevier (In Production).
20. Iqbal, F., & Jha, S. (2018). Closed loop ball end magnetorheological finishing using in-situ roughness metrology. Experimental Techniques, 42(6), 659–669.
21. Iqbal, F., & Jha, S. (2019). Experimental investigations into transient roughness reduction in ball-end magneto-rheological finishing process. Materials and Manufacturing Processes, 34(2), 224–231.
22. Kordonski, W. I. (2003). U.S. Patent No. 6,561,874. Washington, DC: U.S. Patent and Trademark Office.
23. Kordonski, W. I., Shorey, A. B., & Tricard, M. (2004, January). Magnetorheological (MR) jet finishing technology. In ASME International Mechanical Engineering Congress and Exposition (Vol. 47098, pp. 77–84).
24. Kordonski, W. I., Shorey, A. B., & Tricard, M. (2006). Magnetorheological jet (MR Jet TM) finishing technology.

25. Kordonski, W., Shorey, A. B., & Sekeres, A. (2003, December). New magnetically assisted finishing method: material removal with magnetorheological fluid jet. In Optical Manufacturing and Testing V (Vol. 5180, pp. 107–114). International Society for Optics and Photonics.
26. Tricard, M., Kordonski, W. I., Shorey, A. B., & Evans, C. (2006). Magnetorheological jet finishing of conformal, freeform and steep concave optics. CIRP Annals, 55(1), 309–312.
27. DeGroote, J. E., Marino, A. E., Wilson, J. P., Bishop, A. L., Lambropoulos, J. C., & Jacobs, S. D. (2007). Removal rate model for magnetorheological finishing of glass. Applied Optics, 46(32), 7927–7941.
28. Miao, C., Shafrir, S. N., Lambropoulos, J. C., Mici, J., & Jacobs, S. D. (2009). Shear stress in magnetorheological finishing for glasses. Applied Optics, 48(13), 2585–2594.
29. Miao, C., Lambropoulos, J. C., & Jacobs, S. D. (2010). Process parameter effects on material removal in magnetorheological finishing of borosilicate glass. Applied Optics, 49(10), 1951–1963.
30. Arrasmith, S. R., Jacobs, S. D., Lambropoulos, J. C., Maltsev, A., Golini, D., & Kordonski, W. I. (2001, December). Use of magnetorheological finishing (MRF) to relieve residual stress and subsurface damage on lapped semiconductor silicon wafers. In Optical Manufacturing and Testing IV (Vol. 4451, pp. 286–294). International Society for Optics and Photonics.
31. Salzman, S., Romanofsky, H. J., Clara, Y. I., Giannechini, L. J., West, G. J., Lambropoulos, J. C., & Jacobs, S. D. (2013, October). Magnetorheological finishing with chemically modified fluids for studying material removal of single-crystal ZnS. In Optifab 2013 (Vol. 8884, pp. 888407). International Society for Optics and Photonics.
32. Beier, M., Scheiding, S., Gebhardt, A., Loose, R., Risse, S., Eberhardt, R., & Tünnermann, A. (2013, October). Fabrication of high precision metallic freeform mirrors with magnetorheological finishing (MRF). In Optifab 2013 (Vol. 8884, pp. 88840S). International Society for Optics and Photonics.
33. Bedi, T. S., & Singh, A. K. (2018). A new magnetorheological finishing process for ferromagnetic cylindrical honed surfaces. Materials and Manufacturing Processes, 33(11), 1141–1149.
34. Grover, V., & Singh, A. K. (2017). A novel magnetorheological honing process for nano-finishing of variable cylindrical internal surfaces. Materials and Manufacturing Processes, 32(5), 573–580.
35. Sadiq, A., & Shunmugam, M. S. (2009). Investigation into magnetorheological abrasive honing (MRAH). International Journal of Machine Tools and Manufacture, 49(7–8), 554–560.
36. Singh, G., Singh, A. K., & Garg, P. (2017). Development of magnetorheological finishing process for external cylindrical surfaces. Materials and Manufacturing Processes, 32(5), 581–588.
37. Alam, Z., Iqbal, F., Ganesan, S., & Jha, S. (2019). Nanofinishing of 3D surfaces by automated five-axis CNC ball end magnetorheological finishing machine using customized controller. The International Journal of Advanced Manufacturing Technology, 100(5–8), 1031–1042.
38. Saraswathamma, K., Jha, S., & Rao, P. V. (2015). Experimental investigation into ball end magnetorheological finishing of silicon. Precision Engineering, 42, 218–223.
39. Singh, A. K., Jha, S., & Pandey, P. M. (2012). Nanofinishing of fused silica glass using ball-end magnetorheological finishing tool. Materials and Manufacturing Processes, 27(10), 1139–1144.
40. Khan, D. A., Kumar, J., & Jha, S. (2016). Magneto-rheological nano-finishing of polycarbonate. International Journal of Precision Technology, 6(2), 89–100.

41. Khan, D. A., & Jha, S. (2018). Synthesis of polishing fluid and novel approach for nano-finishing of copper using ball-end magnetorheological finishing process. Materials and Manufacturing Processes, 33(11), 1150–1159.
42. Khan, D. A., & Jha, S. (2019). Selection of optimum polishing fluid composition for ball end magnetorheological finishing (BEMRF) of copper. The International Journal of Advanced Manufacturing Technology, 100(5–8), 1093–1103.
43. Alam, Z., Khan, D. A., & Jha, S. (2019). MR fluid-based novel finishing process for nonplanar copper mirrors. The International Journal of Advanced Manufacturing Technology, 101(1–4), 995–1006.
44. Khan, D. A., Alam, Z., Iqbal, F., & Jha, S. (2017). A study on the effect of polishing fluid composition in ball end magnetorheological finishing of aluminum. In 39th International MATADOR Conference on Advanced Manufacturing. UK: University of Manchester.

4 Magnetorheological Abrasive Flow Finishing

4.1 MAGNETORHEOLOGICAL ABRASIVE FLOW FINISHING

The magnetorheological abrasive flow finishing (MRAFF) process is derived from a combination of two processes: magnetorheological finishing (MRF) and abrasive flow machining (AFM). It combines the individual strength of both MRF [1] and AFM [2] processes and utilizes it for controlled finishing of intricate geometries. The AFM process is based on pressurized flow of a polymeric base medium containing abrasives through the intricate shape/geometries of the workpiece [2]. Since the abrading force in AFM process depends upon the viscosity of the base medium, it is quite challenging to control the forces during the machining/finishing operation. This leads to the lack of determinism in control of finishing forces that AFM process is associated with [3]. At the time of development of the MRAFF process, the application of MRF process was limited to simple and some specific geometries [4]. However, being a magnetic field-assisted finishing process, the MRF process possessed the capability to control the finishing forces and thus proved more deterministic with respect to in-process control of finishing forces. Therefore, a hybrid finishing process called the magnetorheological abrasive flow finishing (MRAFF) process was developed by Jha and Jain [5] in the early part of first decade of the 21st century that combined in a single process the versatility and determinism of AFM process and MRF process, respectively, as shown in Figure 4.1. In this way, both versatility of finishing complex geometries and adding determinism by controlling the rheological properties of finishing medium is simultaneously achieved by the MRAFF process.

4.1.1 Mechanism of Material Removal in MRAFF Process

The MRAFF process is also based on magnetorheological polishing (MRP) fluid. In MRAFF process, the arrangement of workpiece and fixture forms a passage through which the magnetically energized/stiffened MRP fluid is extruded in alternating back and forth motion. Out of the entire workpiece surface area, finishing occurs only on the selective portion of workpiece surface that comes under the exposure of the external magnetic field. The rest of the workpiece area remains unaffected by the passage or rubbing of the extruded MRP fluid.

Figure 4.2(a) shows the schematic of material removal mechanisms in MRAFF process. The externally placed electromagnet forms two opposite magnetic poles on either side of the workpiece and fixture arrangement. This makes the magnetic field lines cross through the workpiece and fixture arrangement and form a magnetically energized region (finishing zone) in the extrusion passage. The MRP fluid passing

DOI: 10.1201/9781003228776-4

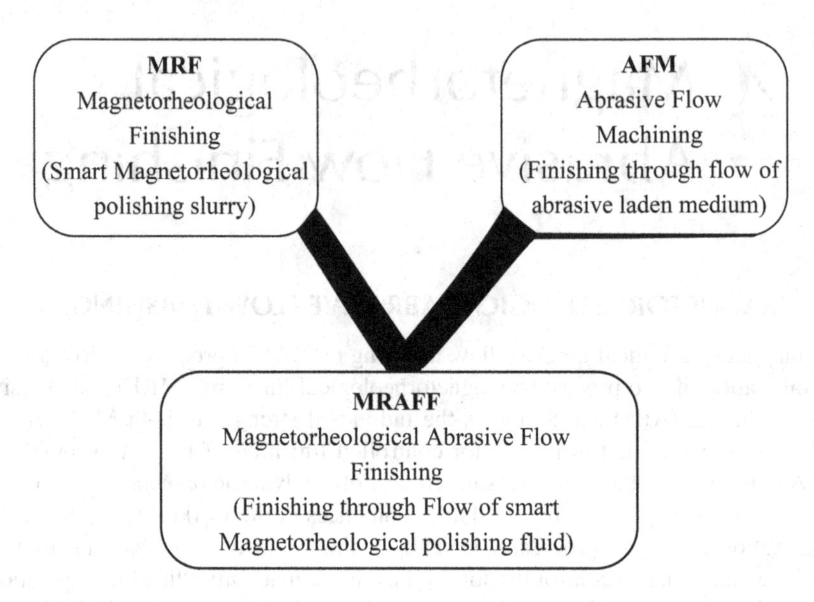

FIGURE 4.1 Magnetorheological abrasive flow finishing process.

through this magnetically energized region gets energized, and its behavior changes from being a Newtonian fluid to a Bingham plastic and again to a Newtonian fluid when it crosses the finishing zone (Figure 4.2b). In the finishing zone or in the Bingham plastic state, the constituents of the MRP fluid align themselves in such a

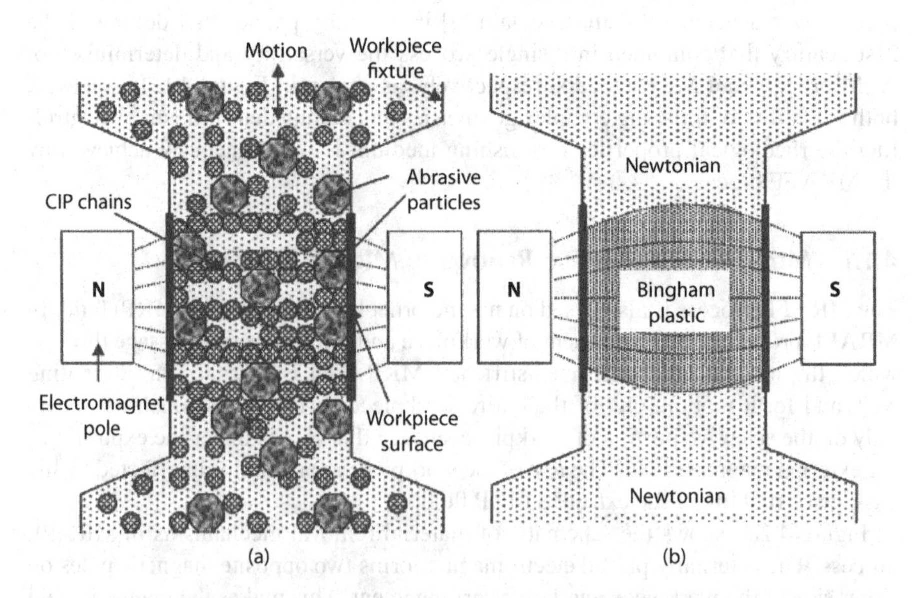

FIGURE 4.2 Schematic of (a) mechanism of material removal in magnetorheological abrasive flow finishing process and (b) magnetorheological polishing fluid's change in rheological behavior in the finishing zone. (From ref. [5].)

way that ferromagnetic particles attach to one another in the direction of magnetic field lines, and the abrasive particles are gripped between these ferromagnetic particle's chains. When the gripped abrasive particle brushes the surface of the workpiece, it breaks the roughness peaks in the form of nano chips, which are carried away by the MRP fluid. There are two factors that govern the quantity of material removed by the abrasive particle brushing the workpiece surface; one is bonding strength of the ferromagnetic particles of MRP fluid and the other is the extrusion pressure.

4.1.2 Experimental Setup

Figure 4.3 shows the schematic of a conventional MRAFF setup developed by Jha and Jain [5]. As seen in the figure, the experimental setup consists of two hydraulic cylinders, two cylinders with piston arrangement for storing the MRP fluid, drives and control unit for operating the hydraulic system, an electromagnet, a workpiece fixture, and support structures.

The two hydraulic cylinders present at the top and bottom are two hydraulic actuators that drive the vertically opposed pistons in the MRP fluid cylinders. The hydraulic cylinders are driven and controlled by the hydraulic drive mechanism that not only imparts reciprocating motion to the cylinders but also ensures that the desired hydraulic pressure is maintained. A variable delivery vane pump driven by a three-phase AC motor is employed for maintaining a constant pressure. The hydraulic drive mechanism is also responsible for correction of any variation in the synchronization of reciprocating strokes by making use of a replenishing circuit that makes up for the loss of oil due to leakage.

The role of MRP fluid cylinders, connected to the top and bottom hydraulic cylinders, is to contain MRP fluid at required extrusion pressure and act as a guide to the reciprocating movement of the piston inside it. The pistons are specially designed and lined with Teflon rings to create the required hydraulic pressure and to prevent any leakage of the MRP fluid. In one of the reciprocating strokes, one piston extends and pushes the MRP fluid from the MRPF cylinder. The pressurized MRP fluid passes through the workpiece and gets collected at the other MRP fluid cylinder where its piston retracts. In the other reciprocating stroke, the reverse happens. In this way, the two pistons work together to push the MRP fluid from one MRP fluid cylinder to another through the workpiece.

The electromagnet arrangement consisting of coils and poles, and the workpiece fixtures are present in between the two MRP fluid cylinders. The workpiece fixture is made up of non-magnetic material and is designed to hold the workpiece and form a passage for the flow of pressurized MRP fluid through it. As shown in Figure 4.3, the workpiece fixture is an assembly of three elements–one cylindrical piece to hold the workpiece(s) and two guiding attachments at each ends of the cylindrical workpiece holder. These guiding attachments facilitate the entry of the pressurized MRP fluid into the cylindrical workpiece holder.

The electromagnet is used to magnetically energize and stiffen the MRP fluid passing through the workpiece. A DC power supply is used to energize the electromagnet coils, and two opposite poles are formed at either ends of the workpiece fixture.

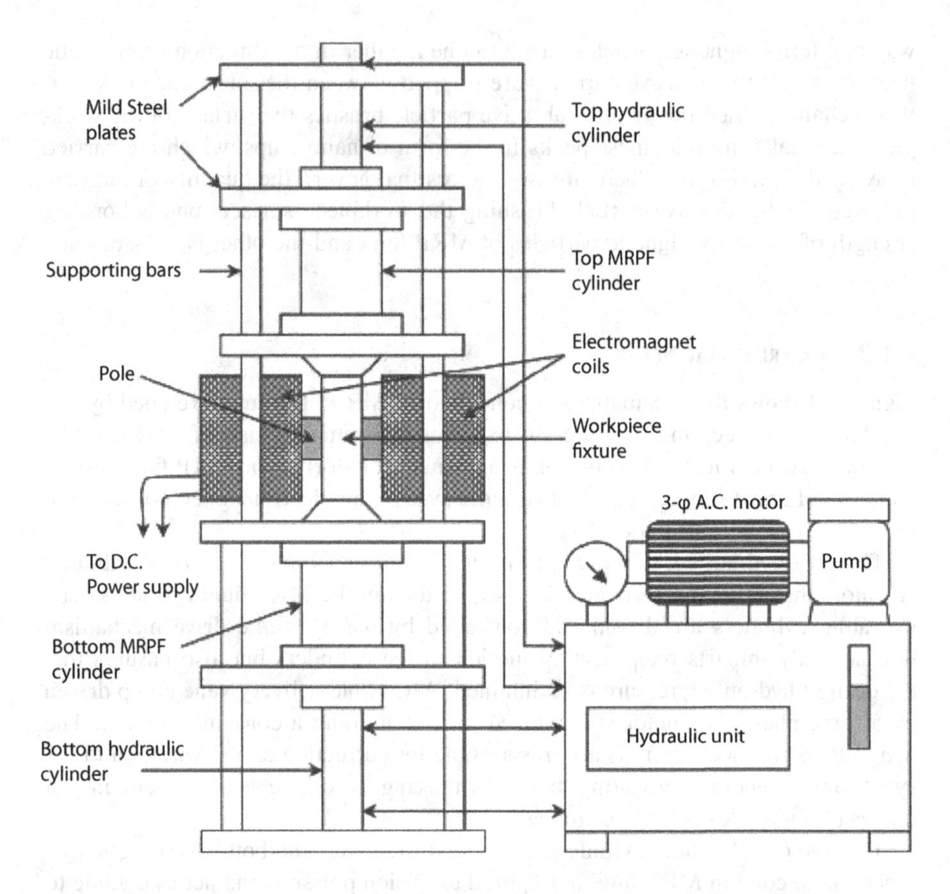

FIGURE 4.3 Schematic of magnetorheological abrasive flow finishing experimental setup. (From ref. [5].)

A certain amount of gap is kept between the two poles of electromagnets to accommodate workpieces of different sizes.

4.2 PROCESS PARAMETERS OF MRAFF

As MRAFF is a magnetic field-assisted finishing process that draws heavily upon electromagnets to impart determinism to its performance, magnetic flux density in the finishing zone plays a major role in influencing the process. The other parameters on which the MRAFF process depends are extrusion pressure, number of finishing cycles, and relative size of carbonyl iron particles (CIP) and abrasive particles used in the MRP fluid.

4.2.1 Magnetic Flux Density

The requirement of an external magnetic field is a must for the MRAFF process as it helps in making the process more deterministic in nature. This makes the magnetic

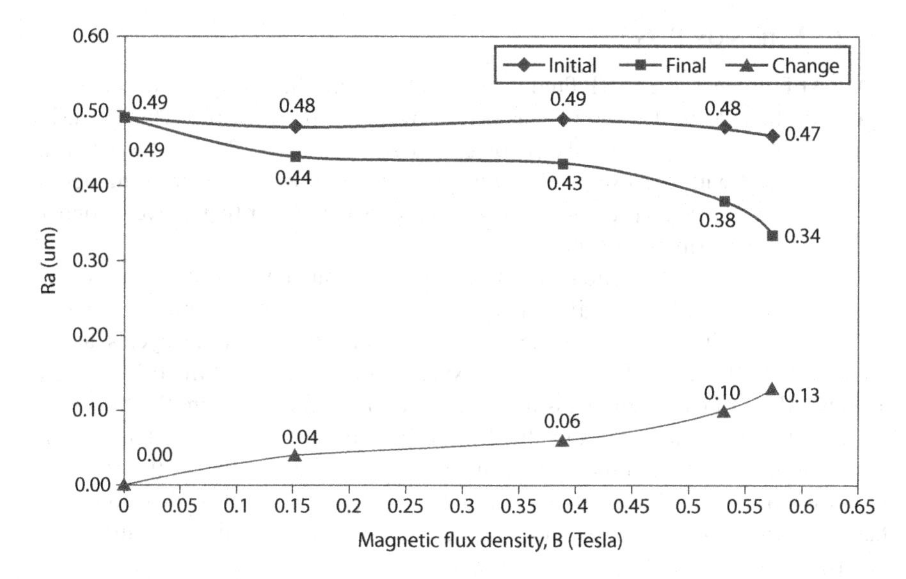

FIGURE 4.4 Effect of magnetic flux density on surface roughness (Ra) value at 3.75 MPa extrusion pressure and 200 finishing cycles. (From ref. [5].)

flux density as one of the most prominent process parameters. The magnetic flux density required in the finishing zone is achieved by using an electromagnet. Powered by a DC supply, the electromagnet is used to control the magnetic flux density in the finishing zone by altering the magnitude of current supplied to the electromagnet windings.

Jha and Jain [5] have studied the effect of magnetic flux density on final surface roughness (Ra) achieved. For an extrusion pressure of 3.75 MPa and 200 finishing cycles, they conducted five experiments. The first experiment was performed without energizing the electromagnet, i.e., at magnetic flux density of 0 T, and the next four experiments were performed at increasing values of magnetic flux density, i.e., at 0.1521, 0.388, 0.531, and 0.574 T. The experimental results obtained by them at various values of magnetic flux density are shown in Figure 4.4. The final surface roughness is reduced with the increase in the magnetic flux density.

In the absence of a magnetic field, no finishing action is observed. This is because in the absence of any magnetic force, the CIPs are incapable of holding the abrasive particles. Hence, when the abrasive particles encounter a roughness peak in their path, they simply roll over and move forward without causing any cutting action. Therefore, the final surface roughness value is same as that of the initial value even after 200 cycles of finishing operations.

A drop in the final surface roughness value is encountered at higher magnetic flux density. This is because high magnetic field leads to increase in the dipole moment between two CIPs, which causes them to stick or bond together strongly [6]. This results in increase in the strength of the CIPs chain links, and hence the CIPs can strongly grip the abrasive particles. As a result of this, high shearing of roughness peaks is encountered at a higher magnetic flux density leading to the decrease in the final surface roughness value of the workpiece.

4.2.2 Extrusion Pressure

In MRAFF process, the MRP fluid is extruded through the workpiece fixture's passage with the help of the two pistons driven by hydraulic actuators. This extrusion of the MRP fluid, along with the rheological transition of the fluid in the finishing zone, achieved with the use of electromagnet, contributes to the finishing action. Therefore, the extrusion pressure is another process parameter that actively controls the finishing performance of the process.

Jha et al. [7] have studied the effect of extrusion pressure on final surface roughness (Ra) achieved during finishing of stainless-steel workpieces at a constant magnetic flux density of 0.53 T and keeping the finishing cycles to 100 numbers for all the experiments. The experimental results obtained by them at various values of extrusion pressure are shown in Figure 4.5. From the figure it is observed that the change in Ra value (initial Ra – final Ra) is the least at 3.0 MPa of extrusion pressure and the maximum at 3.75 MPa. For the rest of the extrusion pressure values, the change in Ra is almost the same. From this, it is concluded that an optimum value of extrusion pressure is required to achieve the maximum material removal.

The pressurized MRP fluid is not solely responsible for material removal. In the workpiece fixture, when the extruded MRP fluid passes through the finishing zone, it is energized by the magnetic field produced by the electromagnet and the material removal in the finishing zone takes place due to the combination of both pressurized and magnetically energized MRP fluid. Due to the presence of magnetic field, the rheological behavior of the MRP fluid changes from Newtonian to a Bingham plastic. In this state of MRP fluid, two modes of abrasion are responsible for material removal depending on the magnitude of field-induced yield stress and applied shear stress.

Assuming the flow of MRP fluid inside the workpiece fixture to be laminar, fully developed, and steady incompressible flow, Jha et al. [7] reported a plug-like unsheared core flowing in the middle of the fixture where the applied stress (τ_{rz})

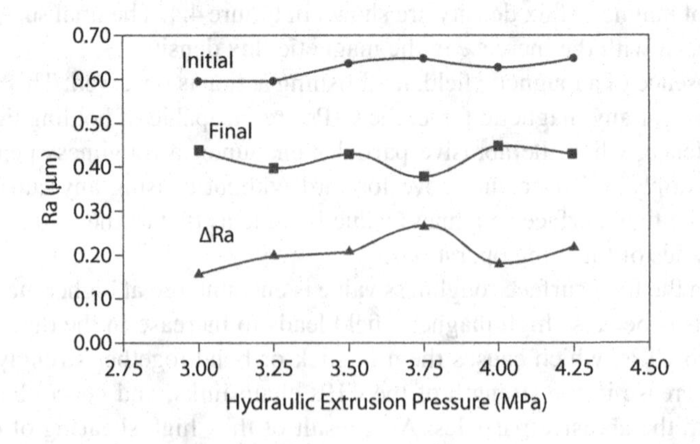

FIGURE 4.5 Effect of hydraulic extrusion pressure on surface roughness (Ra) value at magnetic flux density of 0.53 T and 100 finishing cycles.

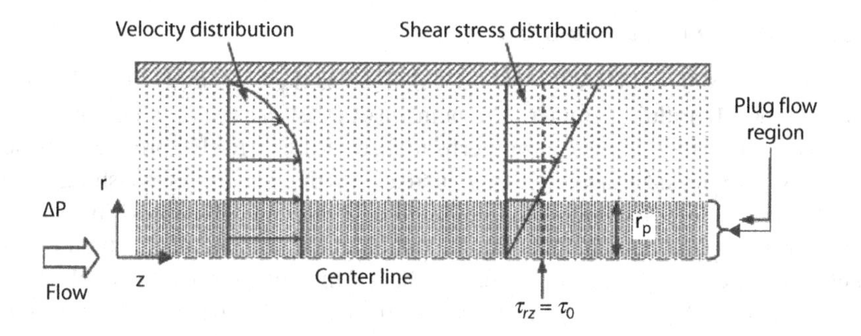

FIGURE 4.6 Schematic of flow of magnetorheological polishing fluid inside the workpiece fixture in Bingham plastic state. (From ref. [7].)

is less than the induced yield stresses (τ_o) as shown in Figure 4.6. The plug flow region's radius is dependent upon wall shear stress and induced yield stress's magnitude. Therefore, the extrusion pressure and the strength of magnetic field control the plug radius r_p (Figure 4.6).

In cases when the plug radius is almost the size of the workpiece fixture's internal radius, the abrasive particles present on the surface of the unsheared core participate in the cutting of the roughness peaks. This type of material removal is attributed to two-body abrasion (Figure 4.7a). For the same magnitude of magnetic field, when the extrusion pressure is increased the MRP fluid forms a sheared layer between workpiece surface and the unsheared core as shown in Figure 4.7(b). In this case, three-body abrasion is encountered where the abrasive particles slide and roll to perform the finishing action [8]. For an increase in the extrusion pressure, the sheared layer increases and greater three-body abrasion is encountered, resulting in the increase in ΔRa (change in Ra). However, beyond an optimum value, if the extrusion pressure is increased further, it leads to a decrease in the plug radius. This results in the vanishing of the three-body abrasive mode, and hence, beyond an optimum pressure, the improvement in the surface finish is not significant.

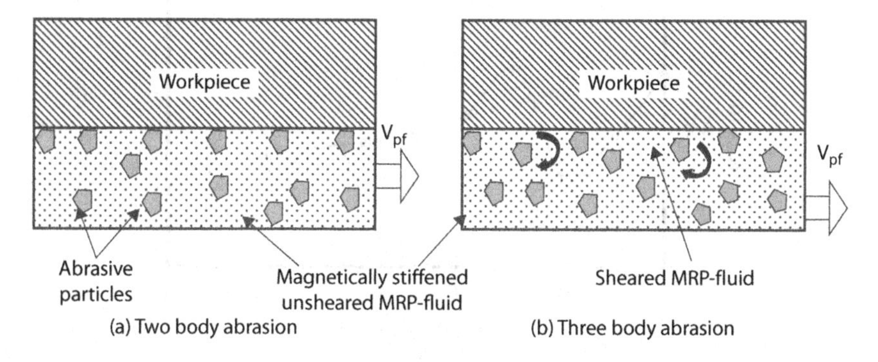

FIGURE 4.7 Modes of abrasive wear in magnetorheological abrasive flow finishing process (a) two-body abrasion and (b) three-body abrasion. (From ref. [7].)

4.2.3 Number of Finishing Cycles

One finishing cycle in MRAFF process is defined as the up-and-down stroke of the piston in the MRP fluid cylinder. The number of finishing cycles is one of process parameters on which the finishing performance of the MRAFF process depends. Generally, one would assume that an increase in the number of finishing cycles would lead to a decrease in the final surface roughness value. However, Jha et al. [7] have reported an interesting phenomenon of "illusive polishing" during their experimentation, where with increase in the number of finishing cycles, the final surface roughness value decreased and then increased substantially before finally decreasing. For an extrusion pressure of 3.75 MPa and magnetic flux density of 0.53 T, they measured the surface roughness after every 100 finishing cycles initially and then after 200 finishing cycles up to 1000 finishing cycles as shown in Figure 4.8.

The surface roughness value decreased till 200 finishing cycles and then increased substantially at the end of 300 finishing cycles. The initial decrease in the first 200 cycles is primarily due to removal of loosely held material left after ploughing during the surface grinding operation. In the next 100 cycles of finishing (i.e., after 300 finishing cycles), the actual deep grinding marks are exposed, and this leads to the substantial increase the surface roughness value. Finishing further (beyond 300 finishing cycles) results in actual finishing operation. Beyond this point there is a gradual decrease in the final surface roughness value. Increasing the number of finishing cycles results in a greater number of abrasives brushing the workpiece surface; hence the roughness peaks are greatly smoothened to result in the lower roughness values.

The phenomenon of "illusive polishing" is important in the sense that most of the magnetic field-assisted finishing processes are superfinishing processes and are carried out after the grinding process and may experience this phenomenon at the initial stage of finishing.

4.2.4 Relative Size of CIP and Abrasive Particles

The size of the CIP and abrasive particles present in the MRP fluid plays an important role in determining the improvement of the final surface finish. Jha and Jain

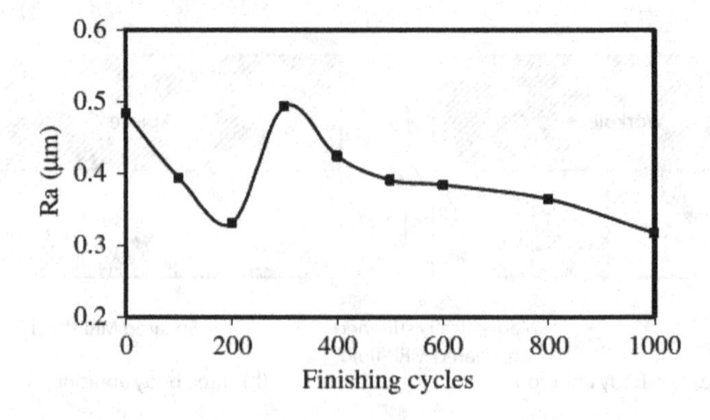

FIGURE 4.8 Effect of number of finishing cycles on surface roughness (Ra) value at magnetic flux density of 0.53 T and extrusion pressure of 3.75 MPa. (From ref. [7].)

TABLE 4.1
Surface Roughness Results for Different Size Combinations of CIP and SiC Abrasive Particles [9]

Exp. No.	CIP Dia (D_{CIP}) (μm)	SiC Dia (D_{SiC}) (μm)	$\frac{D_{CIP}}{D_{SiC}}$	Initial Ra (μm)	Final Ra (μm)	ΔRa[a] (μm)	%ΔRa
1	18.0 (CS)	19.00	0.95	0.32	0.09	−0.23	−71.87
2	18.0 (CS)	12.67	1.42	0.28	0.17	−0.11	−39.28
3	18.0 (CS)	7.50	2.4	0.31	0.23	−0.08	−25.80
4	3.5 (HS)	19.00	0.18	0.26	0.23	−0.03	−11.54
5	3.5 (HS)	12.67	0.28	0.28	0.24	−0.04	−14.28
6	3.5 (HS)	7.50	0.47	0.25	0.24	−0.01	−4.0

[a] ΔRa = final Ra − initial Ra.

[9] studied the role of the relative size of CIPs and abrasive particles on percentage changes in surface roughness (%Δ Ra) in MRAFF process. Two grades of CIP having average particle diameter of 18 μm (CS grade) and 3.5 μm (HS grade) were combined separately with three grades of SiC abrasive particle (800, 1200, and 2000 mesh size) having average particle diameters of 19, 12.67, and 7.5 μm. The percentage reductions in surface roughness for all the six combinations are shown in Table 4.1.

The maximum %Δ Ra is achieved with a combination of CIP CS grade and 800 mesh size SiC abrasive particle. As hypothesized by Jha and Jain [9], the reason behind this is that the chains formed by smaller-sized CIP particles (HS grade) are weak in nature, and they lack the strength to grip the abrasive particles. Another reason hypothesized by them for this behavior is dependent on the magnetic force on the mass of the CIP particles. This is shown in the chain structure formation and unit cell modeling of all the six combinations depicted in Figure 4.9.

4.3 MODELING AND SIMULATION OF MRAFF PROCESS

To have an improved understanding of the material removal mechanism, Jha and Jain [9] modeled the finishing forces involved in the MRAFF process. Using the mathematical expressions for the finishing forces, they also simulated the surface roughness to predict the final surface roughness value for stainless steel workpieces. The simulation depended upon the number of finishing passes and showed the updated roughness value after each finishing pass.

In MRAFF process, the external magnetic field provided by the electromagnets around the workpiece fixture makes the magnetic flux density one of the most prominent functions to be modeled mathematically for finding the finishing forces. Hence, first they [9] established a model for the variation of magnetic flux density between the two electromagnet poles. By considering the two electromagnet poles to be 30 mm apart and assuming each electromagnet a finite solenoid [10] the value of magnetic flux density B at a distance x from one of the electromagnet coils (Figure 4.10),

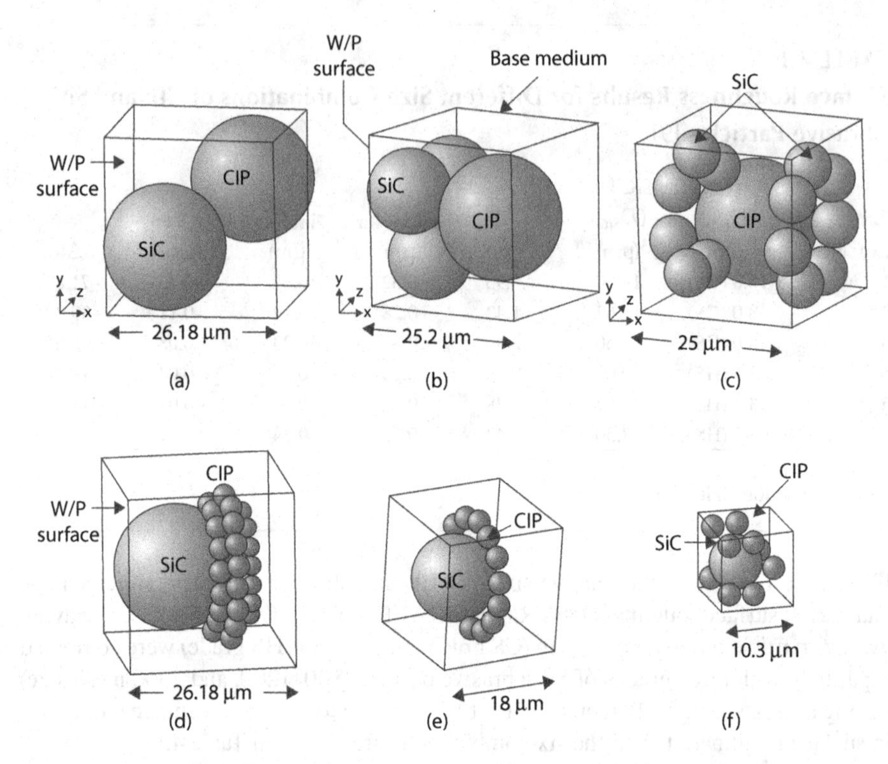

FIGURE 4.9 Configuration of CIPs and SiC abrasives for six different magnetorheological abrasive flow finishing experiments. (a) CIP-CS (18 μm), SiC-800 (19 μm); (b) CIP-CS (18 μm), SiC-1200 (12.67 μm); (c) CIP-CS (18 μm), SiC-2000 (7.5 μm); (d) CIP-HS (3.5 μm), SiC-800 (19 μm); (e) CIP-HS (3.5 μm), SiC-1200 (12.67 μm); (f) CIP-HS (3.5 μm), SiC-2000 (7.5 μm). (From ref. [9].)

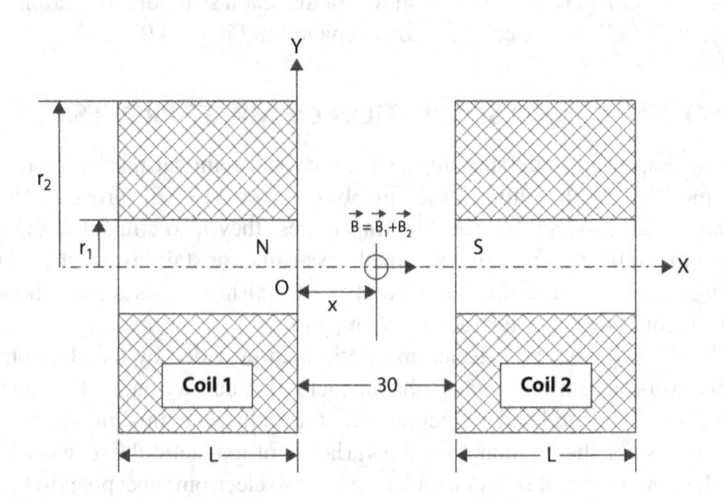

FIGURE 4.10 Configuration of electromagnet for calculation of magnetic flux density (From ref. [9].)

is the vector sum of B_1 (flux density due to coil 1) and B_2 (flux density due to coil 2) and is given by:

$$\vec{B}(\mathrm{x}) = \vec{B}_1(\mathrm{x}) + \vec{B}_2(30 - \mathrm{x}) \tag{4.1}$$

where,

$$\vec{B}_1(\mathrm{x}) = \frac{\mu_0 In}{2(r_2 - r_1)}\left[(L+x)ln\frac{\sqrt{r_2^{\,2}+(L+x)^2}+r_2}{\sqrt{r_1^{\,2}+(L+x)^2}+r_1} - xln\frac{\sqrt{r_2^{\,2}+x^2}+r_2}{\sqrt{r_1^{\,2}+x^2}+r_1}\right] \tag{4.2}$$

$$\vec{B}_1(30-x) = \frac{\mu_0 In}{2(r_2 - r_1)}$$

$$\left[(L+30-x)ln\frac{\sqrt{r_2^{\,2}+(L+30-x)^2}+r_2}{\sqrt{r_1^{\,2}+(L+30-x)^2}+r_1} - (30-x)ln\frac{\sqrt{r_2^{\,2}+(30-x)^2}+r_2}{\sqrt{r_1^{\,2}+(30-x)^2}+r_1}\right] \tag{4.3}$$

where L represents the length of the electromagnet coil, and r_1 and r_2 represent the coil's core and outer radius, respectively. μ_0 is the free space's magnetic permeability; the magnitude of current supplied to the electromagnet is denoted by I; the number of coil-turns per unit length is denoted by n; and x is the distance from the face of the electromagnet pole as shown in Figure 4.10.

After obtaining the expression for the magnetic flux density in the finishing zone, Jha and Jain [9] expressed the magnetic force acting on the CIP in terms of magnetic flux density as:

$$F_m(x) = \frac{m\chi_m}{\mu_o} B(x)\frac{dB(x)}{dx} \tag{4.4}$$

where m and χ_m are CIP's mass and magnetic susceptibility, respectively. This magnetic force acting on the magnetic CIP particle is transferred to the abrasive particle, which is employed during the finishing action. From the expression of the magnetic force that is transferred to the abrasive particle, it is seen that it depends not only on the magnitude of the flux density but also on the gradient of the flux density in the finishing zone.

In MRAFF process, while the magnetic force transferred to the abrasive particle is responsible for penetrating the roughness peaks on the workpiece surface, it is the shear force that is responsible for dislodging the roughness peaks and completing the finishing action. The shear force (F_{shear}) acting on the abrasive particle in MRAFF process, as proposed by Jha and Jain [9], is given by:

$$F_{shear} = (A - A')\tau_y \tag{4.5}$$

where A and A' are the projected area of the abrasive particle and the indented portion of the abrasive in the workpiece surface, respectively, and τ_y is MRP fluid's yield stress.

For simulating the surface roughness in MRAFF process, Jha and Jain [9] developed a program using C programming language. The input surface roughness value to this simulation program was obtained experimentally using a profilometer. After every stroke the simulation updates the surface profile by deducting the depth of indentation (d) of a single spherical abrasive particle on each peak. Subjected to finishing forces, the depth of indentation (d) of a single spherical abrasive particle is given by:

$$d = \frac{D_g}{2} - \frac{1}{2}\sqrt{D_g^{\ 2} - D_i^{\ 2}} \tag{4.6}$$

where D_g and D_i are abrasive grain diameter and indentation diameter, respectively. Further the indentation diameter can be calculated from Brinell hardness number (kgf/mm²) [11] using the following equation given by:

$$H_{BHN} = \frac{2F}{\pi D_g \left(D_g - \sqrt{D_g^{\ 2} - D_i^{\ 2}} \right)} \tag{4.7}$$

where F is the normal force acting on the abrasive particle in N, and H_{BHN} is Brinell hardness number of the workpiece in (kgf/mm²).

Therefore, if the initial heights and depths on the workpiece's surface is given by Y_a, Y_b, Y_c, Y_d, and Y_e as shown in Figure 4.11 then after cutting operation by a single abrasive particle the updated peak heights will be Y_a, $Y_b' = Y_b - d$, $Y_c' = Y_c - d$, Y_d and $Y_e' = Y_e - d$.

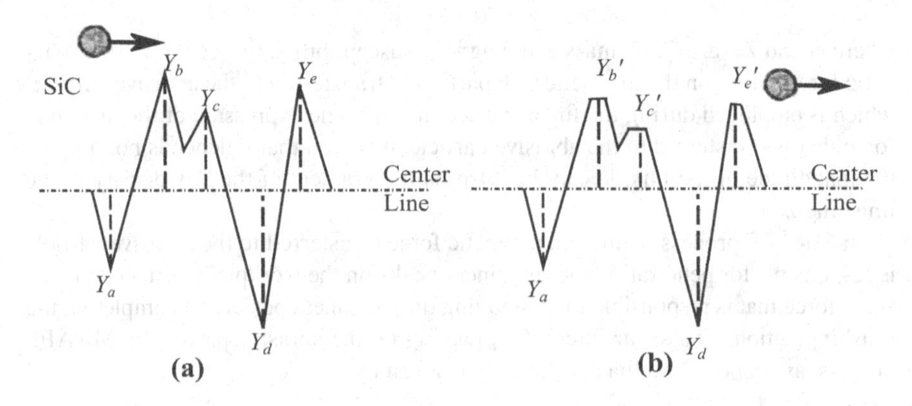

FIGURE 4.11 Schematic showing (a) initial peaks and valleys and (b) updated peaks and valleys after one indentation depth (d). (From ref. [9].)

If the number of active abrasive particles participating in each stroke is denoted by N_g, then the updated peak heights (Y_i) after every stroke is given by:

$$Y_i' = Y_i - N_g d \tag{4.8}$$

The center line average surface roughness (Ra) values is calculated from these peak and valley points using the following equation:

$$R_a = \frac{\sum_{i=1,\ldots N} |Y_i|}{N} \tag{4.9}$$

Das et al. [12–14] performed the CFD simulation of MRP fluid flow inside the workpiece fixture in the MRAFF process. They analyzed the flow of energized fluid in the finishing zone by finite different methods, assuming the fluid to behave like a Bingham plastic. To represent the fluid flow mathematically, they used continuity and momentum equations with the assumptions that the energized MRP fluid is homogeneous, isotropic, and incompressible, and its flow in the finishing zone is steady, fully developed, and axi-symmetric. The equation is given by:

$$-\frac{dp}{dz} + \frac{\mu}{r}\frac{dv_z}{dr} + \mu\frac{d^2v_z}{dr^2} + \left(\frac{d\mu}{dr}\right)\left(\frac{dv_z}{dr}\right) + \rho g_z = 0 \tag{4.10}$$

where v_z and v_z are z and r components of the velocity. $\partial P/\partial z$ is pressure gradient in the z direction; μ is the viscosity.

4.4 ROTATIONAL MRAFF

Rotational MRAFF (R-MRAFF) is a variant of MRAFF process in which the constant magnetic field achieved by a stationary electromagnet placed around the workpiece fixture is replaced by a rotary magnetic field achieved by moving (rotating) permanent magnets [15]. This change is brought about in the region of the workpiece fixture while the rest of the experimental setup resembles that of the conventional MRAFF setup as shown in Figure 4.12(a). The workpiece fixture along with top and bottom MRP fluid cylinders and hydraulic cylinders and the hydraulic drive mechanism are common to both MRAFF and R-MRAFF processes [16].

Several permanent magnets are stacked vertically in a circle and housed in the magnet fixture as shown in Figure 4.12(b). The magnet fixture is made up on non-magnetic material. The fixture is a specially designed cylindrical part that has peripheral slots having dimensions similar to that of the permanent magnets. The slots hold the permanent magnets. The magnet fixture is fixed on the flat top surface of a pulley which, through a timing belt, is connected to the shaft of an AC motor. The rotation of the magnet fixture is achieved through the AC motor and the timing belt arrangement. The RPM of the magnets is controlled by the rotational speed of the motor, which can be altered using a variable frequency drive (VFD). Figure 4.12(c) shows the formation of MRP fluid brush on the inner surface of the fixture due to the presence of permanent magnets.

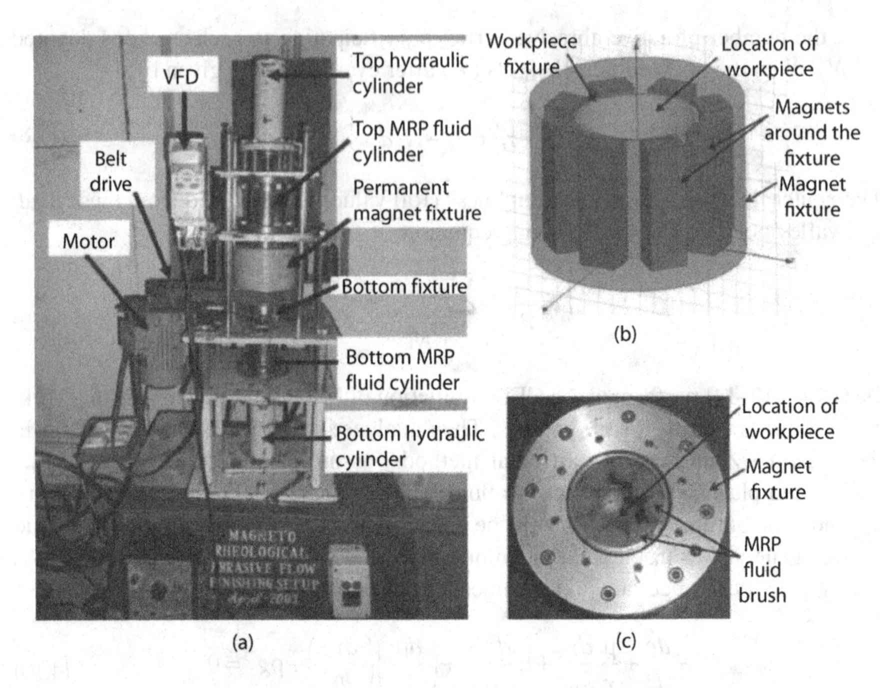

FIGURE 4.12 (a) Photograph of rotational magnetorheological abrasive flow finishing setup, (b) CAD model showing the arrangement f permanent magnets that can be rotated, and (c) formation of MRP fluid brush. (From ref. [16].)

4.4.1 Mechanism of Material Removal in R-MRAFF Process

The working principle or the material removal mechanism of R-MRAFF process is slightly different from that of MRAFF process. In a conventional MRAFF process, the surface is finished by back-and-forth extrusion of magnetically energized MRP fluid. Therefore, the abrasives brush the finishing surface in a straight line while passing through the finishing zone. However, in case of R-MRAFF process, when the energized MRP fluid passes through the finishing zone (Figure 4.13a), it brushes the finishing surface in a helical path (Figure 4.13c), thus increasing the interaction between the abrasive and workpiece surface over a longer length in a single stroke [17]. This increases the finishing efficiency and substantially reduces the finishing time. The helical path followed by the abrasive particle is due to the resultant force (F_c) arising from the axial force (F_a) and the tangential cutting force (F_t) as shown in Figure 4.13(b). The axial force is due to the reciprocation of the medium at velocity, V_a and the tangential cutting force is caused by the rotation of the permanent magnet fixture, which in turn rotates the MRP fluid.

4.4.2 Process Parameters of R-MRAFF

Since the R-MRAFF process is based on the MRAFF process, the process parameters of MRAFF viz. magnetic flux density extrusion pressure and number of finishing cycles are also the same for the R-MRAFF process. However, the magnetic

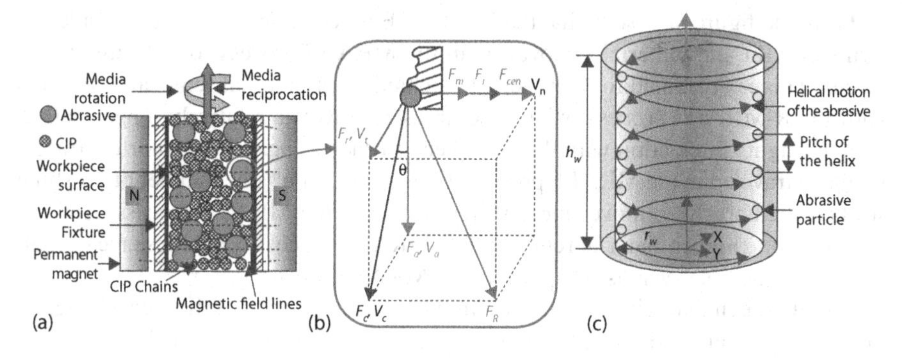

FIGURE 4.13 (a) Schematic showing finishing zone of rotational magnetorheological abrasive flow finishing process, (b) forces and velocity components acting on abrasive, and (c) helical finishing path of an abrasive particle. (From ref. [17].)

flux density, which, in MRAFF process, can be varied by altering the magnetizing current requires change of permanent magnets to achieve the same in the case of R-MRAFF process. The effects of these parameters have already been discussed for the MRAFF process, and their effects on R-MRAFF are more or less the same.

Apart from these parameters, the rotational speed of the permanent magnet is one parameter that is unique to the R-MRAFF process. Das et al. [17] have studied the effect of rotational speed of a magnet on ΔRa and material removal. For an extrusion pressure of 37.5 bar and 600 finishing cycles, the conducted experiments were performed on brass, stainless steel, and EN-8 workpieces at 0, 20, 40, 60, 80, and 100 RPM of rotational speed permanent magnet. At 0 RPM, the R-MRAFF process behaves like an MRAFF process; hence the results as shown in Figure 4.14 serve as a good comparison between the two processes apart from studying the effect of rotational speed.

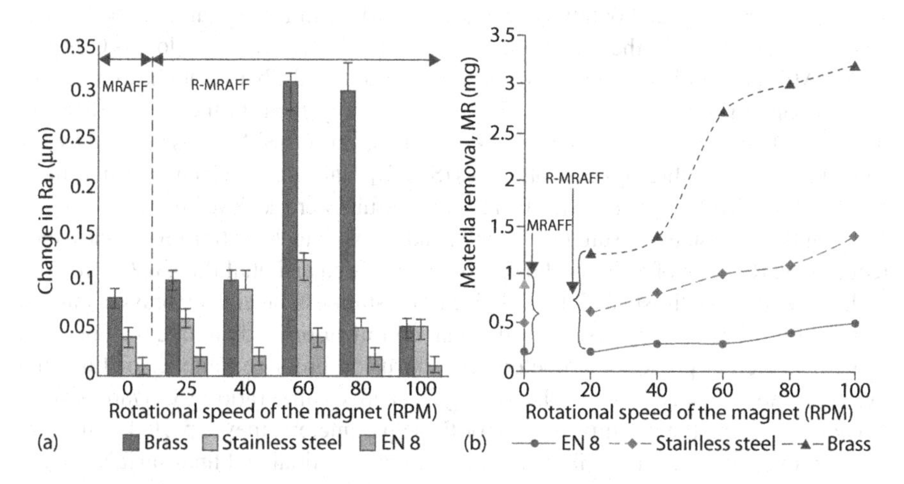

FIGURE 4.14 Effect of rotational speed of magnet on (a) ΔRa and (b) material removal for brass stainless steel and EN-8 workpieces at 37.5 bar extrusion pressure and 600 finishing cycles. (From ref. [17].)

From the figure, it is seen that the R-MRAFF process yields far better results (in terms of ΔRa and material removal) than the MRAFF process for all three workpieces. This is attributed to the inherent mechanism of material removal associated with the R-MRAFF process, where the abrasive travels a relatively longer helical path during its interaction with the workpiece surface as against the shorter straight line path traveled in the MRAFF process. In this way, for a same number of finishing strokes, the abrasive removes more material than in the MRAFF process. Another reason for the higher material removal in R-MRAFF process is the presence of multiple magnetic poles versus two poles in the MRAFF process.

It is also seen that with the increase in the rotational speed of the electromagnet, material removal for the three samples increases. This is because with an increase in the rotational speed, the helix angle of the helical finishing path as shown in Figure 4.13(c) reduces, and the abrasive travels a longer finishing path and removes more material in a single stroke.

4.5 APPLICATIONS

MRAFF and R-MRAFF processes have been employed to finish a variety of materials like brass, stainless steel, EN-8 engineering steel, silicon nitride, materials for knee implants, etc. It has also shown its capability to finish all types of surfaces like planar, non-planar, freeform, internal, and external cylindrical surfaces. Compared to the MRAFF process, R-MRAFF has been more widely used for different finishing applications because of the higher material removal rate associated with it.

Using MRAFF process, Jha and Jain [5] finished flat stainless-steel workpieces and reduced the final surface roughness (Ra) value to 0.34 μm from an initial value of 0.47 μm. They achieved this result at 0.574 T of magnetic flux density and for 200 cycles of finishing. In another example of finishing flat workpieces of stainless steel by MRAFF process, Jha and Jain [9] altered the size of the MRP fluid's constituents (CIP and abrasive particle) and obtained an excellent surface finish on stainless steel. From an initial Ra of 0.32 μm, they were able to achieve a final Ra value to as low as 0.09 μm.

MRAFF process is not only used for finishing of metals. but its application is also seen in nanofinishing of ceramics, which are generally hard, brittle, and difficult to machine. Making use of the MRAFF process, silicon nitride (Si_3N_4) was finished by Jha and Jain [18]. Of the three types of abrasives (SiC, B_4C, and diamond) employed by them for finishing of Si_3N_4, they reported that the best results were achieved by SiC abrasives. Keeping the extrusion pressure to 3.75 MPa and employing 2000 finishing cycles, they reduced the Ra value of Si_3N_4 workpiece to 0.10 μm from an initial Ra of 0.28 μm.

Das et al. [17] finished flat workpieces of stainless steel and brass using an R-MRAFF process. They first carried out an optimization study to find the optimum value of different process parameters like extrusion pressure, number of finishing cycles, rotational speed of the electromagnet. and volume ratio of CIP/abrasive in MRP fluid. Using the optimized value of these parameters, they reported a substantial improvement in terms of finishing rate where the initial and final surface roughness values for stainless steel was Ra = 0.26 μm and Ra = 0.11 μm and for brass was Ra = 0.24 μm and Ra = 0.05 μm. The actual photographs of the initial and finished surfaces of both stainless steel and brass are shown in Figure 4.15.

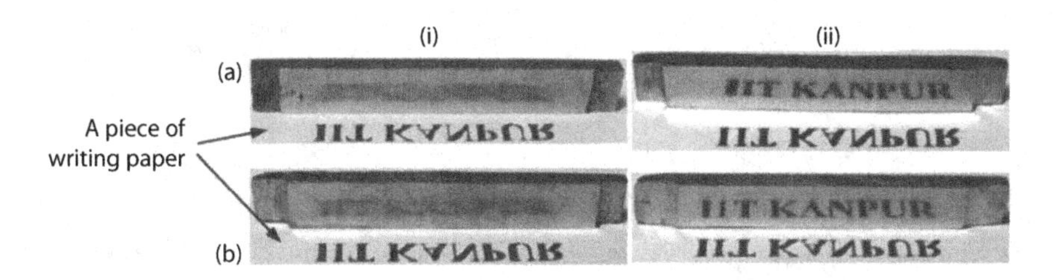

FIGURE 4.15 Photographs of (a) stainless steel workpiece and (b) brass workpiece; (i) initial surface and (ii) after finishing by R-MRAFF process. (From ref. [19].)

In another application of R-MRAFF, Das et al. [19] finished stainless steel tube using SiC as abrasives. In this study too, first, the different process parameters involved were optimized and then the stainless-steel tube was finished using the optimized process parameters. The photograph, roughness profile and AFM image of the initial and finished surface are shown in Figure 4.16. The internal surface of the tube for photograph and measurement is obtained by parting off the tube. Starting with the initial Ra value of 330 nm, they finished the steel tube to obtain the final Ra value of 16 nm. Such low value of surface roughness results in extreme finishing of surfaces and this is evident from the initial and final surface photograph of the workpiece (Figure 4.16a) where the reflection of the letters IITK is not seen in the initial surface of the tube but is clearly reflected on the final finished surface.

The R-MRAFF process is not only capable of finishing the internal surface of tubes but it can also be used to correct the roundness error of cylindrical tubes. Using

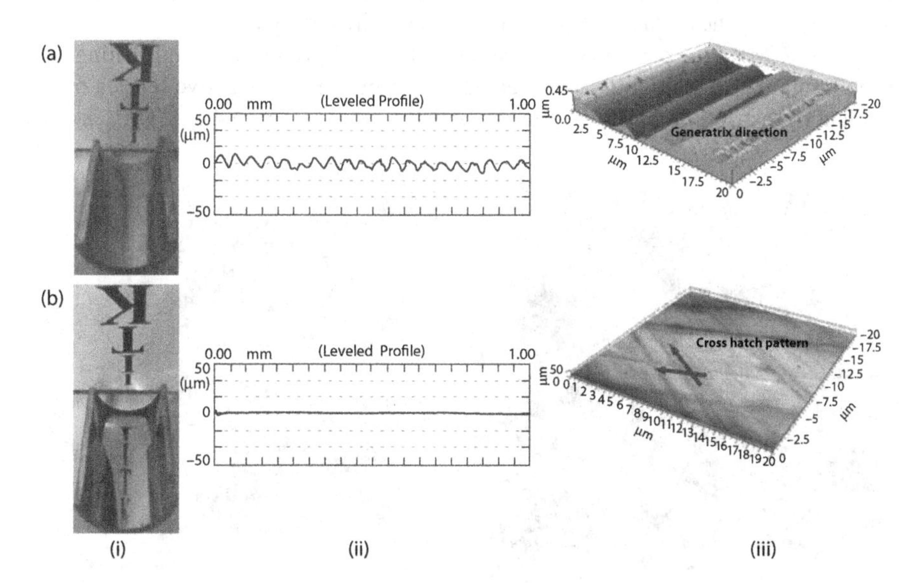

FIGURE 4.16 (i) Photograph of internal workpiece surface after parting off, (ii) surface roughness profile, and (iii) AFM image showing surface texture of (a) initial workpiece surface (Ra = 330 nm), that does not reflect the letters IITK, (b) final workpiece surface (Ra = 16 nm) after finishing by R-MRAFF process, which reflects the letters IITK. (From ref. [19].)

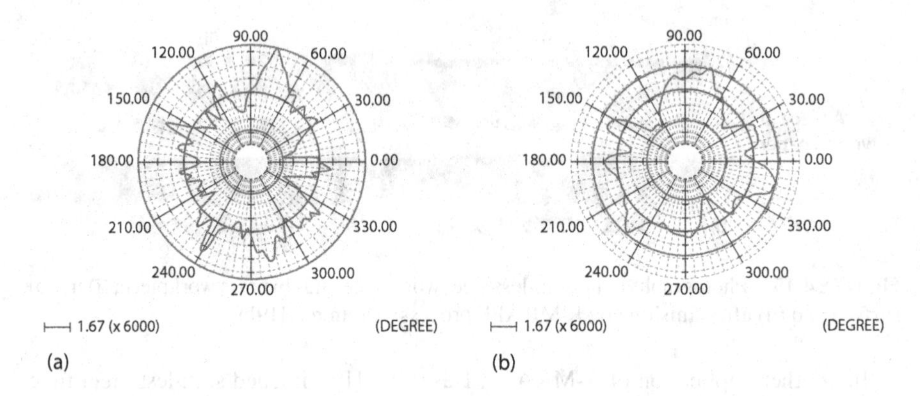

FIGURE 4.17 (a) Out-of-roundness (OOR) profile for unfinished surface (OOR = 5.17 μm) and (b) OOR profile for finished surface (OOR = 3.17 μm). (From ref. [20].)

the R-MRAFF process, Das et al. [20] reduced the out-of-roundness (OOR) of cylindrical stainless-steel components from 5.17 μm to 3.39 μm as seen in Figure 4.17.

Kumar et al. [16] demonstrated the free-form surface-finishing capability of the R-MRAFF process. In the work reported by them, they finished a stainless-steel workpiece similar to the knee joint implant in multiple steps, where in each step, the permanent magnets are rotated to rotate the flowing MRP fluid against the stainless-steel surface mimicking the implant. The result obtained by them is shown in Figure 4.18. Although they reported a decrease in the overall finishing time compared to MRFF (magnetorheological fluid-based finishing) process [21], the value of surface roughness was not uniform over the entire workpiece.

Using R-MRAFF process, Nagdeve et al. [22, 23] proposed nanofinishing of a knee implant with a slightly modified approach by using a negative replica of the implant to restrict the MRP fluid in the polishing zone (Figure 4.19). This ensures

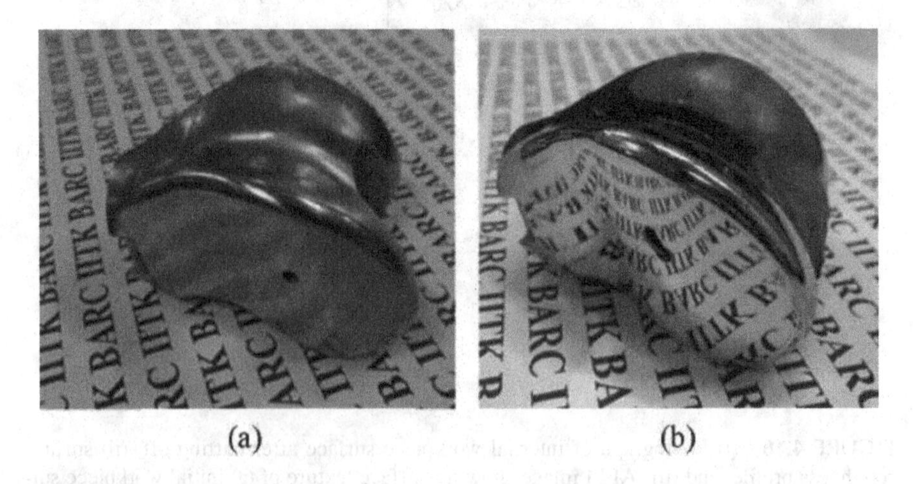

FIGURE 4.18 Photograph of a stainless-steel workpiece similar to the knee joint implant (a) before finishing and (b) after finishing. (From ref. [16].)

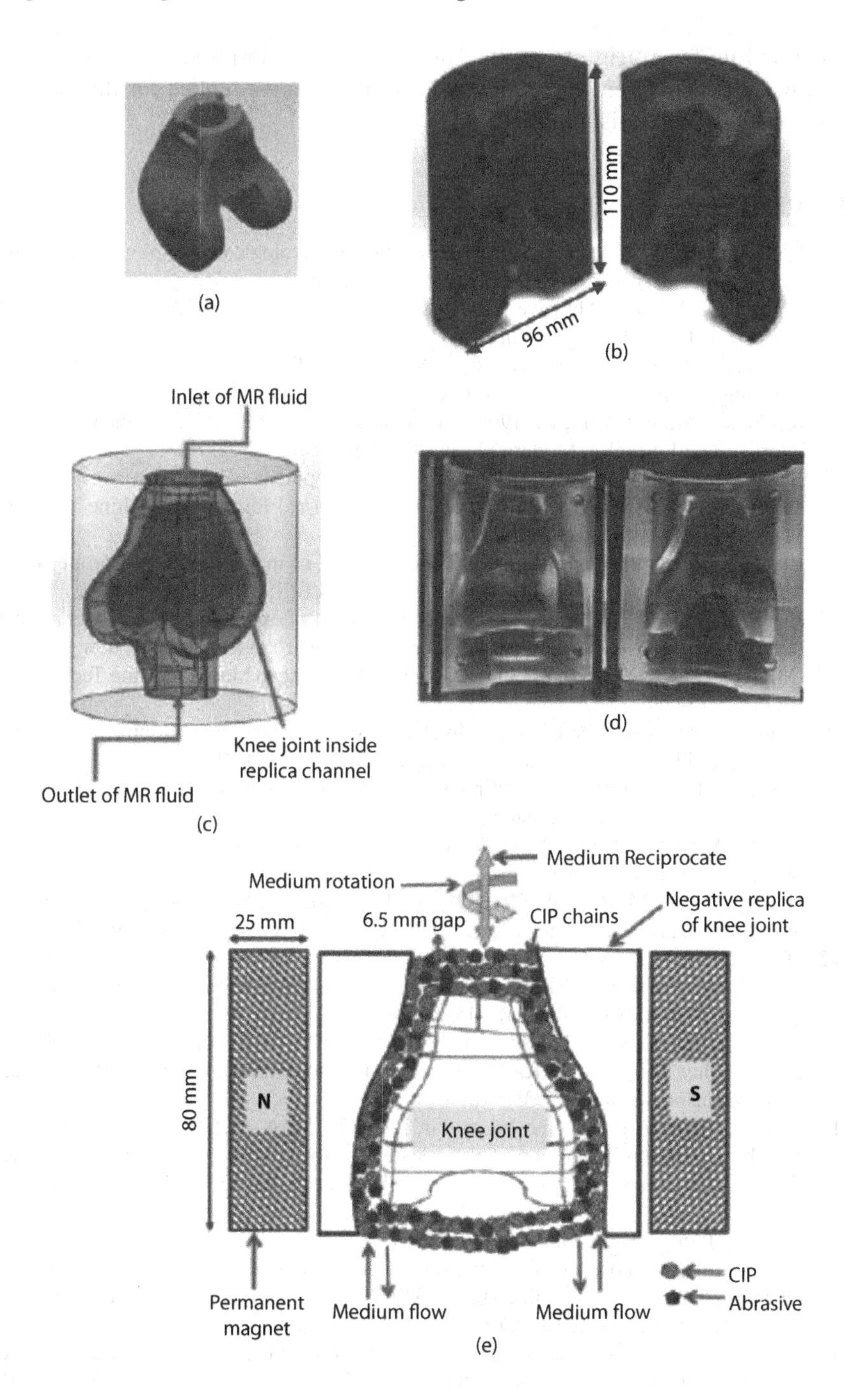

FIGURE 4.19 (a) Solid model of knee joint; (b) Solid model of negative replica of knee joint; (c) Transparent view of knee joint in negative replica; (d) Fabricated replica of knee joint; (e) MRP fluid flowing through the uniform gap between replica and workpiece surface. (From ref. [23].)

a tight and uniform grip of the fluid over the surface to provide uniform finish (Figure 4.19e). However, the requirement of the negative replica for this method adds to the cost and finishing time.

REFERENCES

1. Kordonski, W. I., & Jacobs, S. D. (1996). Magnetorheological finishing. International Journal of Modern Physics B, 10(23n24), 2837–2848.
2. Rhoades, L. (1991). Abrasive flow machining: a case study. Journal of Materials Processing Technology, 28(1–2), 107–116.
3. Jain, V. K. (Ed.). (2016). Nanofinishing Science and Technology: Basic and Advanced Finishing and Polishing Processes. CRC Press.
4. Kordonski, W., & Golini, D. (1999). Progress update in magnetorheological finishing. International Journal of Modern Physics B, 13(14n16), 2205–2212.
5. Jha, S., & Jain, V. K. (2004). Design and development of the magnetorheological abrasive flow finishing (MRAFF) process. International Journal of Machine Tools and Manufacture, 44(10), 1019–1029.
6. Huang, J., Zhang, J. Q., & Liu, J. N. (2005). Effect of magnetic field on properties of MR fluids. International Journal of Modern Physics B, 19(01n03), 597–601.
7. Jha, S., Jain, V. K., & Komanduri, R. (2007). Effect of extrusion pressure and number of finishing cycles on surface roughness in magnetorheological abrasive flow finishing (MRAFF) process. The International Journal of Advanced Manufacturing Technology, 33(7–8), 725–729.
8. Fang, L., Zhou, Q. D., & Li, Y. J. (1991). An explanation of the relation between wear and material hardness in three-body abrasion. Wear, 151(2), 313–321.
9. Jha, S., & Jain, V. (2006). Modeling and simulation of surface roughness in magneto-rheological abrasive flow finishing (MRAFF) process. Wear, 261(7–8), 856–866.
10. Ida, N., & Bastos, J. P. (2013). Electromagnetics and Calculation of Fields. Springer Science & Business Media.
11. Hayden, H. W., Moffatt, W. G., & Wulff, J. (1965). The Structure and Properties of Materials. V. 3. Mechanical Behavior. John Wiley and Sons, New York, p. 248.
12. Das, M., Jain, V. K., & Ghoshdastidar, P. S. (2008). Fluid flow analysis of magnetorheological abrasive flow finishing (MRAFF) process. International Journal of Machine Tools and Manufacture, 48(3–4), 415–426.
13. Das, M., Jain, V. K., & Ghoshdastidar, P. S. (2008). Analysis of magnetorheological abrasive flow finishing (MRAFF) process. The International Journal of Advanced Manufacturing Technology, 38(5–6), 613–621.
14. Das, M., Jain, V. K., & Ghoshdastidar, P. S. (2015). A 2D CFD simulation of MR polishing medium in magnetic field-assisted finishing process using electromagnet. The International Journal of Advanced Manufacturing Technology, 76(1–4), 173–187.
15. Das, M., Jain, V. K., & Ghoshdastidar, P. S. (2009, January). Parametric study of process parameters and characterization of surface texture using rotational-magnetorheological abrasive flow finishing (R-MRAFF) process. In International Manufacturing Science and Engineering Conference (Vol. 43628, pp. 251–260).
16. Kumar, S., Jain, V. K., & Sidpara, A. (2015). Nanofinishing of freeform surfaces (knee joint implant) by rotational-magnetorheological abrasive flow finishing (R-MRAFF) process. Precision Engineering, 42, 165–178.
17. Das, M., Jain, V. K., & Ghoshdastidar, P. S. (2012). Nanofinishing of flat workpieces using rotational–magnetorheological abrasive flow finishing (R-MRAFF) process. The International Journal of Advanced Manufacturing Technology, 62(1–4), 405–420.

18. Jha, S., & Jain, V. K. (2006). Nanofinishing of silicon nitride workpieces using magnetorheological abrasive flow finishing. International Journal of Nanomanufacturing, 1(1), 17–25.
19. Das, M., Jain, V. K., & Ghoshdastidar, P. S. (2010). Nano-finishing of stainless-steel tubes using rotational magnetorheological abrasive flow finishing process. Machining Science and Technology, 14(3), 365–389.
20. Das, M., Jain, V. K., & Ghoshdastidar, P. S. (2011). The out-of-roundness of the internal surfaces of stainless steel tubes finished by the rotational–magnetorheological abrasive flow finishing process. Materials and Manufacturing Processes, 26(8), 1073–1084.
21. Sidpara, A. M., & Jain, V. K. (2012). Nanofinishing of freeform surfaces of prosthetic knee joint implant. Proceedings of the Institution of Mechanical Engineers, Part B: Journal of Engineering Manufacture, 226(11), 1833–1846.
22. Nagdeve, L., Jain, V. K., & Ramkumar, J. (2016). Experimental investigations into nano-finishing of freeform surfaces using negative replica of the knee joint. Procedia CIRP, 42, 793–798.
23. Nagdeve, L., Jain, V. K., & Ramkumar, J. (2018). Differential finishing of freeform surfaces (knee joint) using R-MRAFF process and negative replica of workpiece as a fixture. Machining Science and Technology, 22(4), 671–695.

5 Process Automation of Magnetic Field Assisted Finishing

5.1 INTRODUCTION

Automated mechanical processes are the current need in industrial scenarios. Factory equipment is expected to deliver the highest throughput with minimum possible production cost. In different industries, including oil, gas, and petrochemicals, energy costs can amount to a range of 30–50% of the total production cost. In this view, the magnetic field assisted finishing processes, researched and developed at a brisk pace over the past 2 decades, have reached at a stage where they can be utilized as proper industrial systems.

This, however, does not seem to be the case with current industrial facilities, which still rely on classical finishing methods of manual labor to achieve precise surface finish on parts. A reason for the magnetic field assisted finishing processes not being able to be utilized properly in the industry is lack of proper automation of such processes and systems. To address this lack, this chapter begins by defining "mechanical process automation" as follows:

> Mechanical process automation is defined as monitoring of data from different variables involved in a process, analyzing this monitored data, feeding the analysis to the process controller to make decisions of carefully manipulating process actuators to obtain desired optimum results.

This chapter then goes on to describe process parameters and their role in automation. It further discusses various process and motion parameters for select magnetic field assisted finishing processes. This is followed by a detailed description of the control hardware and control panel design required for an automated mechanical system. Special emphasis has been given to feedback systems and closed-loop control of magnetic field assisted finishing processes. The chapter concludes by discussing a case study of closed-loop control of a magnetic field assisted finishing process viz ball end magnetorheological finishing (BEMRF) process.

5.2 PROCESS PARAMETERS AND THEIR CHARACTERIZATION

The behavior of a mechanical process is characterized by its numerical or other measurable factor forming one of a set that defines a system or sets the conditions of its operation. These factors are termed as parameters, and each magnetic field

DOI: 10.1201/9781003228776-5

assisted finishing process has its own governing parameters which define how the process will react to manipulation of these parameters. The measure of output in a magnetic field assisted finishing process varies according to the manipulation of these parameters.

There are parameters that may affect the performance of a process but are not directly associated with the process; rather they are associated with the geometry of the part being processed. The parameters involved in characterization and performance measure of magnetic field assisted finishing processes are broadly classified into following four categories:

a. **Input parameters:** These are the parameters a user provides to the system; they can be user-controlled in terms of input parameters, such as magnetizing current, spindle speed, working gap, etc.
b. **Output parameters:** The parameters that are a measure of the process performance and cannot directly be manipulated by a user are output parameters, such as surface roughness and material removal rate.
c. **Process parameters:** Parameters that govern the performance of a process directly are process parameters. In each process, these parameters are same irrespective of geometry of a part being processed.
d. **Motion parameters:** These parameters are related to the geometry of a part being processed and hence are different for different parts in the same process.

Process parameters can fit in categories of input as well as output parameters, whereas the motion parameters are largely input parameters for magnetic field assisted finishing processes.

Apart from the four categories defined for the parameters of magnetic field assisted finishing processes, these are also classified based on their automation hardware characteristics. These usually define the control automation hardware for the manipulation of a parameter. The broad classification on this basis is divided into the following two categories:

a. **Digital parameters:** These parameters require a digital signal to be passed on to the controller for their manipulation. They are on and off type only and do not affect the process performance in terms of output but are essential of the process to be carried out properly specially in case of a process being controlled automatically. Examples of digital parameters include enabling and disabling of a power supply for magnetizing current, enabling and disabling of pump for magnetic polishing fluid supply, and control of directional valves in case of hydraulics used as motion control in some processes.
b. **Analog parameters:** These parameters require an analog signal to be passed on to the controller for their manipulation and they directly govern the performance of a process. The control hardware devices required for these parameters are also required to be compatible with analog behavior. Examples of these parameter types include level of magnetizing current, rate of fluid flow from pump, working gap, spindle speed, table feed rate, and proportional speed of hydraulic ram actuation.

The digital and analog parameters are not only required for input parameters but also for output parameters in control of magnetic field assisted finishing processes. This includes control of the roughness measurement system of a process, enabling and disabling of this system, speed control of movement of measurement probe, control of actuators for cleaning systems, and other similar actuations. The automation of magnetic field assisted finishing processes requires many such digital and analog parameters which may or may not directly affect the performance of a process but are essential for proper automation of the process. The detailed discussion on such parameters will be covered in this chapter in forthcoming sections. The following section provides a brief overview of magnetic field assisted finishing processes and the parameters involved with them.

5.2.1 Process Parameters of Magnetic Abrasive Finishing (MAF)

In magnetic abrasive finishing (MAF) [1], normally an electromagnet is employed to convert electrical current into a magnetic field. However, the electromagnet can have an alternative in the form of a permanent magnet to generate a magnetic field. The latter case has one distinct disadvantage of having a fixed magnetic flux, and therefore, effective control of the magnetic force is not achieved; therefore, an electromagnet is always preferred. In this process, iron particles and fine abrasive particles are sintered with each other to form ferromagnetic abrasive particles. An electromagnet is rotated above a workpiece surface with a tiny gap in between the two, which is filled by the ferromagnetic abrasive particles.

The parameters involved in MAF are primarily the working gap, spindle rotating speed, table feed rate, and magnetizing current. The first three are considered as process parameters even though they involve motion aspects in all three of them, but the characterization of motion parameters has been limited to geometry of the part only. There are other parameters like abrasive size etc. but discussion in this chapter will focus on parameters that relate to automation of the process. The MAF process has two derivatives of MAF viz. double disc MAF (DDMAF) [2] and ultrasonic assisted MAF (UAMAF) [3]. The two have already been explained in previous chapters; the parameters involved in DDMAF are similar as in MAF explained previously. In addition to the parameters in DDMAF, the UAMAF has few more parameters which are related to the ultrasonic power supply. The ultrasonic-related process parameters are pulse on-time, pulse off-time, enabling disabling of the power supply. Each of the ultrasonic parameters are digital, whereas overall in MAF both analog and digital parameters are being used. The automation of MAF and its derivatives will require precise control of these parameters in tandem with a feedback system in place. A dedicated section at the end of this chapter will focus on possibilities of how to automate the MAF process and its derivatives.

5.2.2 Process Parameters of Magnetorheological Finishing (MRF)

Like MAF, the MRF process [4] is also governed by magnetic field assistance but with the additional feature of rheological characteristics that help control the forces of finishing. For automation consideration, we will focus on variants of MRF and parameters involved in them. At the outset, the parameters involved in MRF processes are magnetizing current, rotational speed of tool spindle, and working gap. However, with the development of various corollaries of MRF process the parameters associated have also been developed into more than just three basic parameters.

Ball end magnetorheological finishing (BEMRF) [5], magnetorheological jet finishing (MRJF) [6], and magnetorheological abrasive honing (MRAH) [7] are a few variants that have very recently been developed, i.e., in the last decade or so. In the BEMRF process, the automation development has been the most advanced with dedicated work being done to automate the process and even develop fully automated machine tools for the process. The parameters involved in automation of BEMRF are numerous when it comes to a complete set of parameters. Apart from the basic finishing parameters various digital and analog process parameters and motion parameters have been associated with successful automation of the BEMRF process. The automation of the BEMRF process will be discussed as a case study at the end of this chapter to provide detailed aspects of how magnetic field assisted finishing processes are to be automated.

The MRJF process was also developed with automation functionalities at the very beginning of the process development, the parameters involved in automation of MRJF are jet velocity, jet diameter, abrasive size, and concentration. This is one process where it is a must to have automation capabilities well in place to control the process and achieve desired results. The MRJF process is very difficult to control without proper automation in place. Another variant of MRF is the magnetorheological abrasive honing (MRAH) process, this process involves magnetizing a current or magnetic field, rotational speed of the spindle, and frequency of reciprocation of the plunger as the parameters needed to properly carry out the magnetic field assisted finishing. For accuracy in results and increase in production, proper automation of the process is needed; for this the control of parameters is important like all other magnetic field assisted finishing processes described earlier as well. For MRAH, the frequency of reciprocation is a unique parameter and will require very different automation aspects compared to other processes.

5.2.3 Process Parameters of Magnetorheological Abrasive Flow Finishing (MRAFF)

In magnetorheological abrasive flow finishing (MRAFF) [8] process, the role of automation intensifies with unique features required for proper control, which include control of hydraulics as well as pump and electromagnet. The MRAFF does have a variant to go with which is known as rotational MRAFF or R-MRAFF; this was developed to address a few limitations in MRAFF. The process parameters involved in R-MRAFF [9] process are hydraulic pressure of extrusion, frequency of extrusion, rotational speed of the magnet, and fluid volumetric composition. The difference between MRAFF and R-MRAFF is simply an addition of another degree of freedom to the electromagnet motion. As the name suggests, the motion is rotational, and for automation of the process, another level of complexity is added to control R-MRAFF, which makes the process much more capable than MRAFF alone.

5.3 MOTION PARAMETERS AND CONTROL

Each magnetic field assisted finishing process requires control of motion parameters for proper control of movement of workpiece and the tool. With automation being the focus of such processes, the control of motion parameters becomes vital and thus generates a need to understand the motion parameters in general and then specific to each magnetic field assisted finishing process as well. The motion aspects of each magnetic field assisted finishing process require considerable power as the tool and worktable are of considerable weight to lift and move. There is more than one way of providing this power to the worktable and tool movements. These ways are discussed in the following subsections.

5.3.1 Types of Drives and Actuators

For magnetic field assisted finishing processes the motion given to the worktable or tool is imparted by actuators, which in most cases are motors; in some cases, they can be hydraulic or pneumatic actuators as well.

The controller is programmed to run these actuators with desired speed and give them desired direction to move. Running the program means the controller sends command signals to achieve these desired tasks from the actuators.

These command signals given by the controller are for the purpose of information only and do not have enough strength in them to make the actuator move. The drive (Figure 5.1) then takes these command signals as inputs and performs amplification

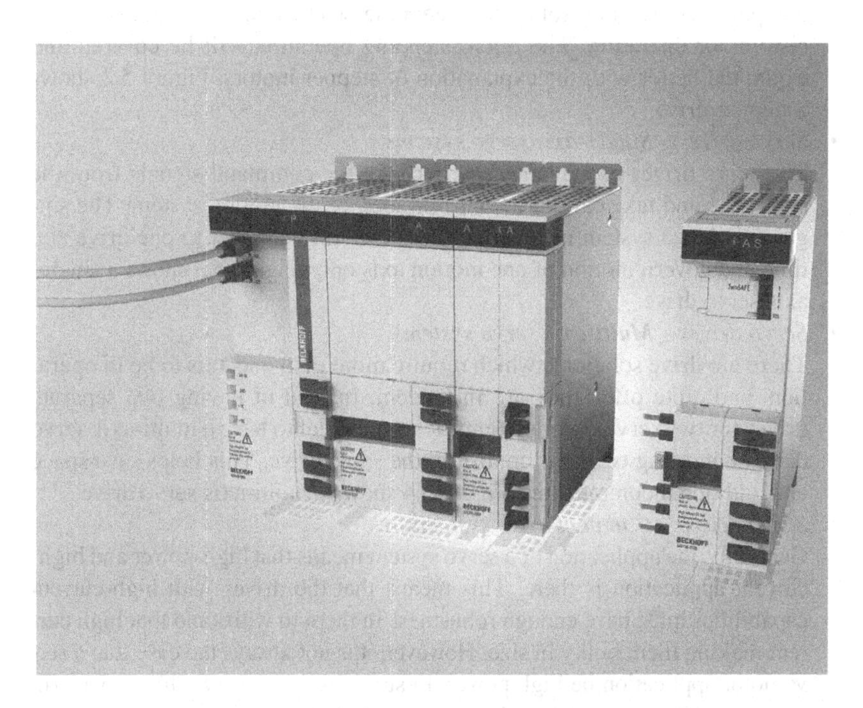

FIGURE 5.1 Servo drive system AX8000 series. (Pic. credits: Beckhoff Automation GmbH & Co. KG.)

on them with proper voltage and current levels required to make an actuator move as desired. Therefore, the drives are also known as amplifiers. In recent advancements, the drives have become smart and perform many of the complex tasks of a controller. These complex and sophisticated functions are handling the motor feedback, and they close the current loop, velocity loop, and the position loop. Each of the axes of motion systems typically operate under closed-loop control. There are three loops that each axis needs to close, viz. current loop, velocity loop, and position loop. The actuator's velocity and position are continuously monitored and provided back to the controller, while their current consumed are monitored and given back to the drive, i.e., the drive closes the current loop.

5.3.1.1 State-of-the-Art Availability of Drives

There are various types of drives available in the market and are categorized based on either the actuator they are controlling or the functionality they can perform. Some of state-of-the-art drive systems currently available are:

- *Stepper drives*
  The stepper drives are intended for control of stepper motors and are generally open-looped systems with no feedback. They provide pulse-width modulated (PWM) output for the motor coils. The current advancement in technology has meant that the majority of drives can be adjusted to a specific stepper motor, and the application is being used just by changing a few parameters. Multi-fold micro-stepping ensures noiseless and very precise motor operation. The micro-stepping operation will be covered and explained better with the explanation of stepper motors. Figure 5.2 shows a stepper drive.
- *Servo drives – Single-axis servo systems*
  The servo drives as explained earlier take the command signals from the controller and take care of the current loop of the motion system. The single-axis servo system means that these can only connect to one drive at a time and govern motion of one motion axis only. Figure 5.3 shows a single-axis servo drive.
- *Servo drives – Multi-axis servo systems*
  There are drive solutions, which require more than one axis to be in operation, and quite often they are in tandem. Instead of having two separate drives for two servomotor systems, it is beneficial to have a multi-axis servo drive supporting two-axis control in the same drive. This helps save space and additional complexities. Figure 5.4 shows a multi-axis servo drive.
- *Servo drives – Compact drive systems*
  Generally, the application of a servo system means that high-power and high-current application is there. This means that the drives with high-current capabilities must have enough robustness in them to withstand that high current making them bulky in size. However, it is not always the case that a servomotor application be high power. In such cases, the servo drives perform the same tasks but can be made compact in size and look like an I/O module. Figure 5.5 shows a compact servo drive with one-cable technology (OCT).

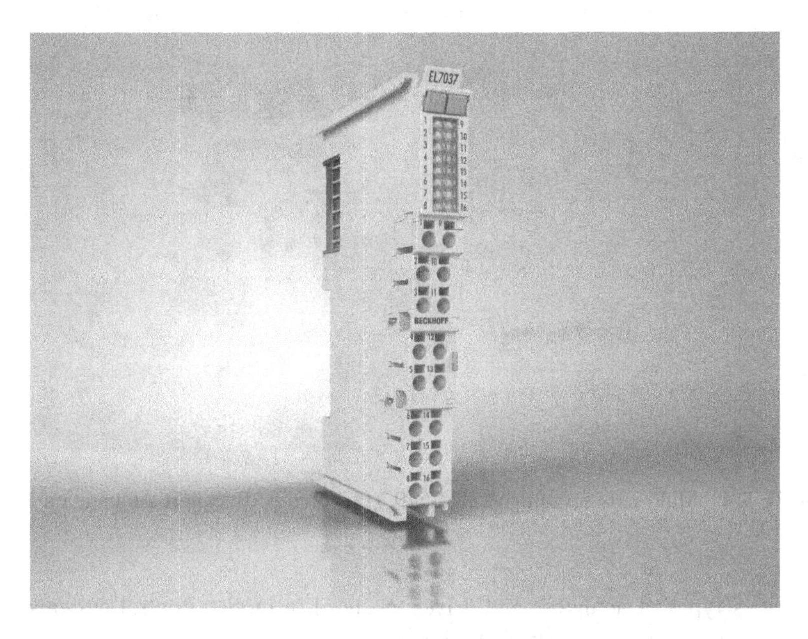

FIGURE 5.2 Stepper drive EL7037. (Pic. credits: Beckhoff Automation GmbH & Co. KG.)

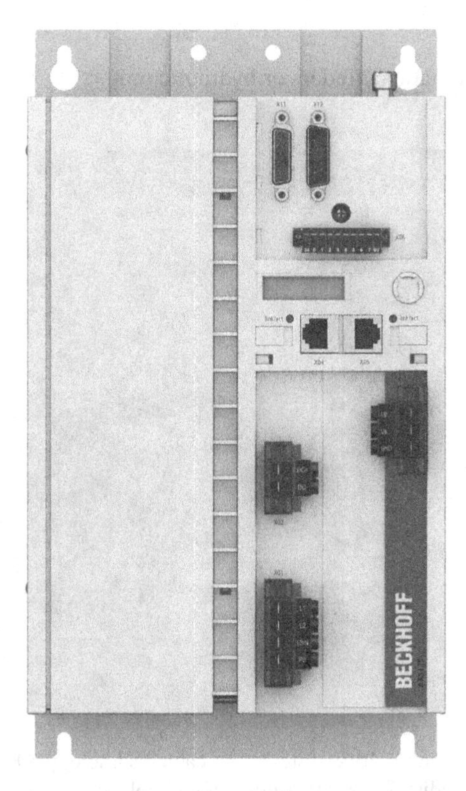

FIGURE 5.3 Single-axis servo drive AX5125. (Pic. credits: Beckhoff Automation GmbH & Co. KG.)

FIGURE 5.4 Multi-axis servo drive AX8540. (Pic. credits: Beckhoff Automation GmbH & Co. KG.)

Various types of actuators and drives are used in motion control systems. The actuator types for motion control systems are:

- Servomotors
- Stepper motors
- Hydraulic/pneumatic cylinders or hydraulic motors

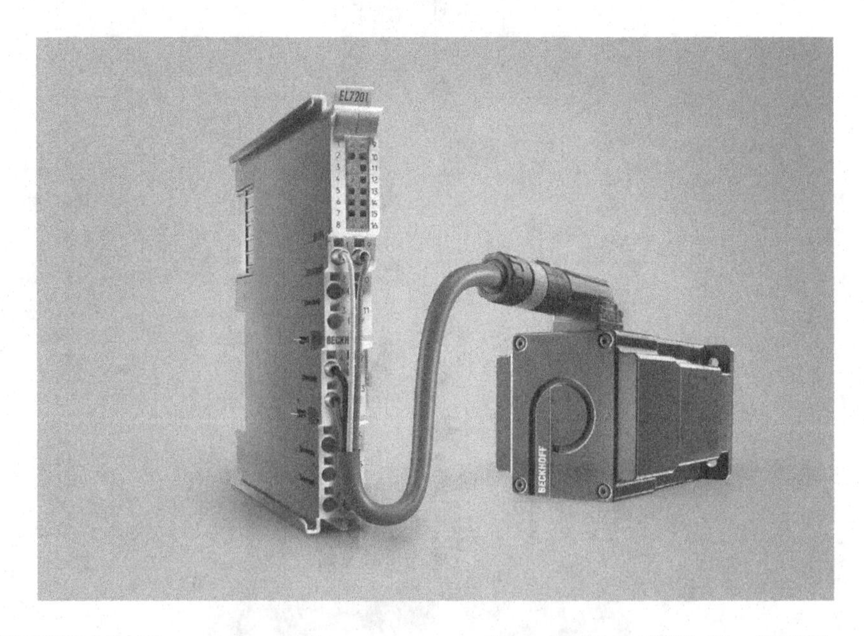

FIGURE 5.5 Compact servo drive system: one-cable technology (OCT) drive EL7201 and motor AM8131. (Pic. credits: Beckhoff Automation GmbH & Co. KG.)

FIGURE 5.6 Servomotor AM8063. (Pic. credits: Beckhoff Automation GmbH & Co. KG.)

The servomotor (Figure 5.6) is a servomechanism, which means it uses position input to control its movement and end position. The servomotor houses an in-built encoder to give position and speed feedback to the drive for a closed-loop operation. In the least efficient scenario, only position is measured and fed back. The deliberate position of the output is contrasted with the charge position, the outer information to the controller. As the positions approach, the error between the current position and the desired position approaches zero and the motor stops when it is zero.

The stepper motor (Figure 5.7), as opposed to the servomotor, is an open-loop actuator but can be controlled precisely. The shaft of a stepper motor rotates in discrete step increments when electrical command pulses are applied to it in the proper sequence. The motor rotation has several direct relationships to these applied input pulses. The sequence of the applied pulses is directly related to the direction of the motor shafts rotation. The speed of the motor shafts' rotation is directly related to the frequency of the input pulses, and the length of rotation is directly related to the number of input pulses applied. One of the most significant advantages of a stepper motor is its ability to be accurately controlled in an open-loop system. Open-loop control means no feedback information about position is needed. This type of control eliminates the need for expensive sensing and feedback devices such as optical encoders. Your position is known simply by keeping track of the input step pulses. There are different ways in which a stepper drive controls a stepper motor, and these are called excitation modes viz. full-step, half-step, and micro-stepping modes.

The third type of actuators, i.e., the hydraulic/pneumatic cylinders, are used where not much position control is required but a to-and-fro motion of a fixed stroke length is desired. These types of actuators (Figures 5.8–5.10) are either run by compressed air known as pneumatic actuators, or they are run by pressurized liquid known as hydraulic cylinders.

Pneumatic actuators are preferred where less power and more speed are desired whereas hydraulic actuators are used where high power is required and speed may

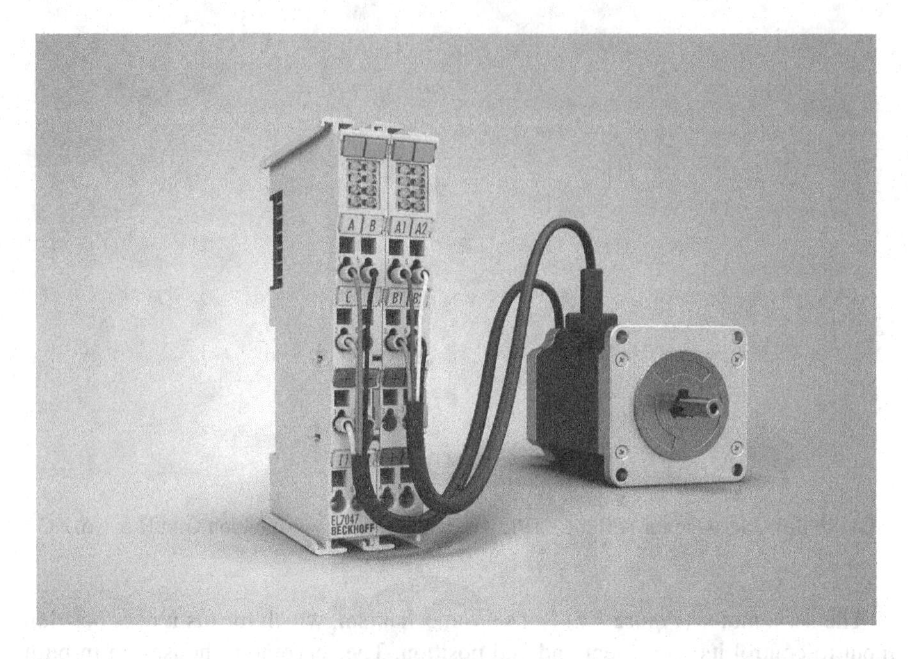

FIGURE 5.7 Stepper motor (AS1010) with drive (EL7047). (Pic. credits: Beckhoff Automation GmbH & Co. KG.)

not be a big concern. The hydraulic motor is used when the use of electrical or electronics is not desired, and the rotary motion requirement is fulfilled using the hydraulic motor.

In case of hydraulic or pneumatic actuators, it must be noted that the analogy to servo or stepper drives are valves. Typically, directional control valves are used to

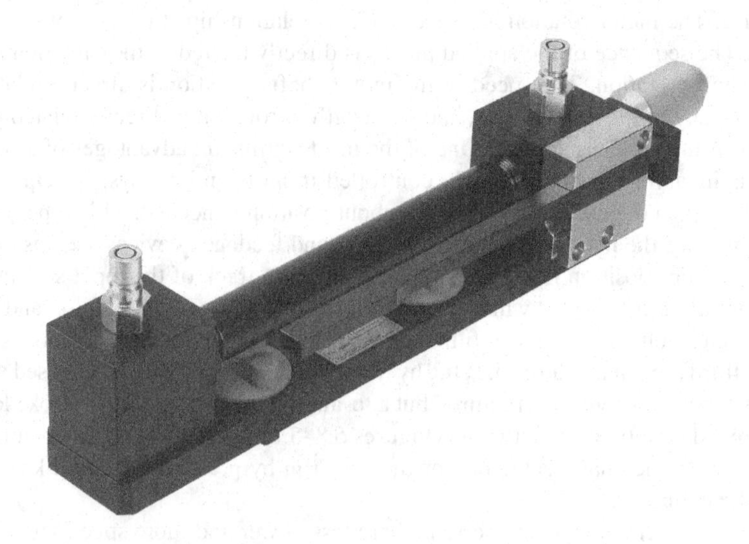

FIGURE 5.8 Hydraulic cylinder. (Pic credit: Festo Didactic SE.)

FIGURE 5.9 Pneumatic cylinder. (Pic credit: Festo Didactic SE.)

control the direction of motion of the actuator, i.e., forward, or reverse stroke length in case of cylinder-type actuators and clockwise or anti-clockwise in case of motor-type actuators.

Other than the directional control valves, there are servo valves (Figures 5.11 and 5.12) and proportional valves, which enable precise control of pressure of the hydraulic fluid enabling speed variation and power variation in the automated system. Later

FIGURE 5.10 Hydraulic motor. (Pic credit: Festo Didactic SE.)

FIGURE 5.11 Directional control valve (DCV). (Pic credit: Festo Didactic SE.)

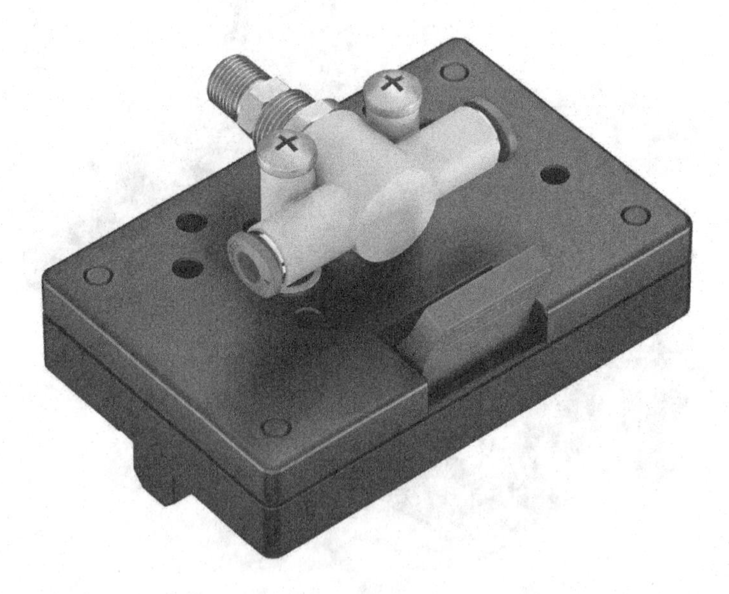

FIGURE 5.12 Proportional throttle valve. (Pic credit: Festo Didactic SE.)

in this chapter, we will see what motion type is usually chosen for each magnetic field assisted finishing process and why.

5.3.2 Number of Axes

The number of axes of motion is the degree of freedom a system can move in. The system here can be either the worktable, the toolhead, or a combination of both in case of magnetic field assisted finishing processes. The movement of this system along these axes can be achieved by any of the actuators mentioned in previous sections in this chapter. Typically, there are six degrees of freedom for a body to move in free space. In manufacturing machines, the motion axes utilized are either three, four, or five, depending upon the requirement for a particular application. The first three axes are always translatory in nature, i.e., linear motion of X, Y, and Z axes of the cartesian coordinate system. The other two are rotary motion axes, which are any two of A, B, and C axes, where A, B, and C are rotations about X, Y, and Z axes, respectively. Figure 5.13 presents a schematic representation of the six motion axes available for motion control of machining processes, a maximum of only five of these are usually employed at a time.

Conventionally the motion axes of a machining system are motors employed to rotate lead screws for linear motion and rotary slides for rotary motion. However, there are cases where hydraulic or pneumatic cylinders are employed for linear motion, and hydraulic motors for rotary motion; the latter is very rarely used.

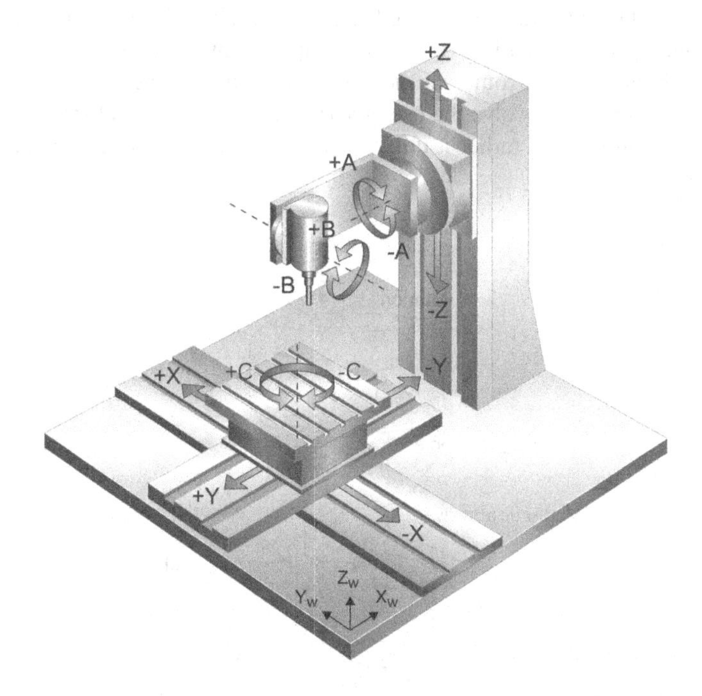

FIGURE 5.13 CNC axes system showing six axes available for motion. (Pic credits: ISG GmbH – [www.isg-stuttgart.de].)

5.3.3 Motion Options in Each Process

In magnetic field assisted finishing processes, there are different motion configurations employed in each process. This section will briefly explain the current motion options available with each such process under consideration. The processes under consideration are the same as discussed earlier in this chapter, and these are viz. MAF (UAMAF, DDMAF), MRF (BEMRF, MRAH, MRJF), and MRAFF along with R-MRAFF.

5.3.3.1 Motion Configuration of MAF and Its Variants

Figure 5.14 shows a schematic representation of the MAF process, which is typically a three-axes motion arrangement. The motion options in it are the X and Y axes motion given to the worktable and the Z axis motion given to the toolhead.

In case of UAMAF, the setup is quite like MAF with an addition of ultrasonic vibrations given to the worktable as schematically shown in Figure 5.15. In case of DDMAF, again the worktable is imparted with linear motion in X and Y directions, whereas the toolhead is given linear motion in Z direction, and two rotating discs assist finishing in DDMAF. One of the discs is attached to the spindle and the other to the workpiece to facilitate finishing action. The DDMAF motion configuration is schematically shown in Figure 5.16.

5.3.3.2 Motion Configuration of MRF and Its Variants

Figure 5.17 shows the schematic representation of motion configuration of MRAH setup. In this, the workpiece is attached to spindle and is given rotation on its own axis, the only other motion in MRAH is the movement of plunger that pushes and pulls the MR fluid to and from over the workpiece. This magnetic field assisted finishing process does not involve the use of servo or stepper motors for the movement of polishing fluid plunger rather it utilizes the hydraulic cylinder as an actuator for this purpose.

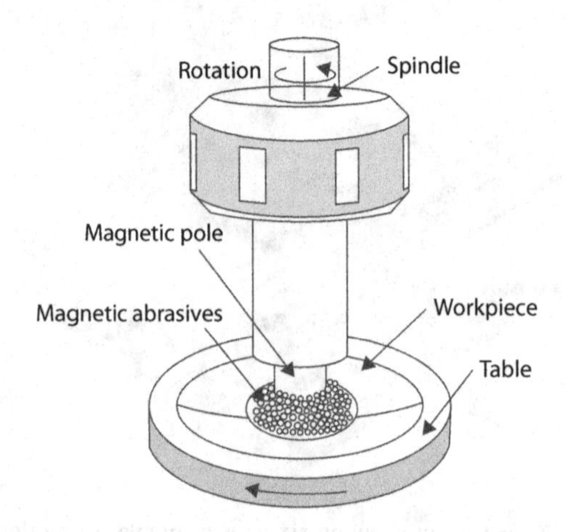

FIGURE 5.14 Schematic view of free-form magnetic abrasive finishing. (From ref. [1].)

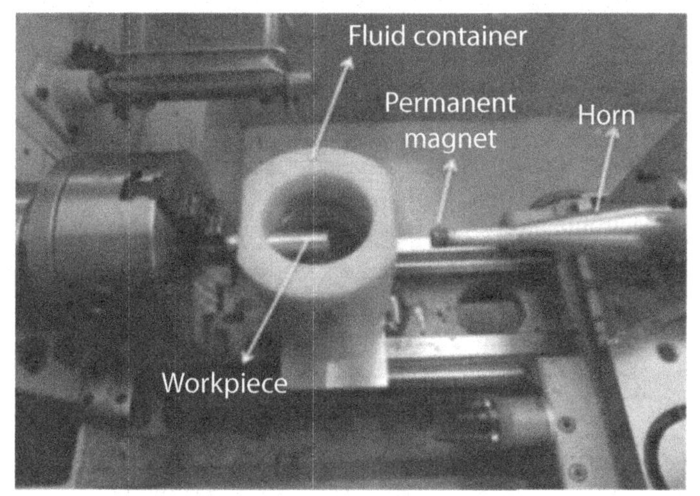

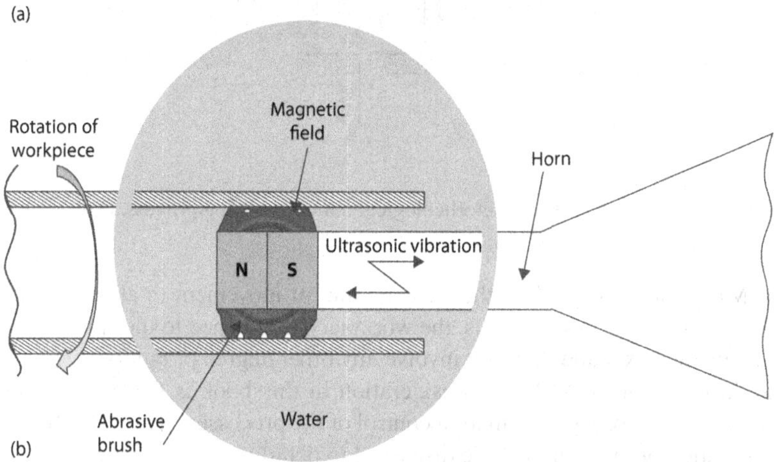

FIGURE 5.15 Ultrasonic assisted magnetic abrasive finishing process: (a) Actual photograph, (b) Schematic illustration. (From ref. [3].)

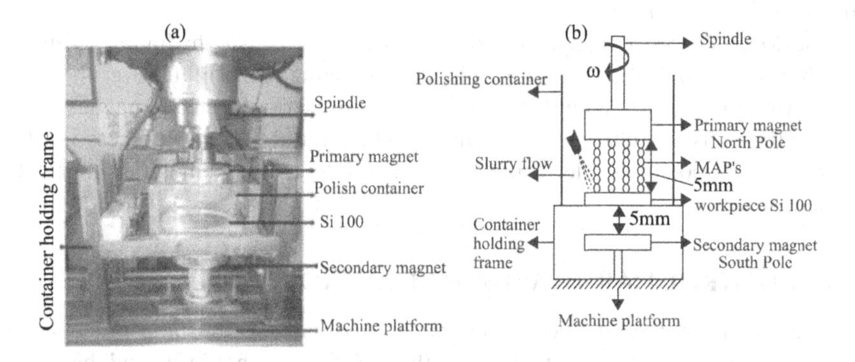

FIGURE 5.16 Double disc magnetic abrasive finishing process: (a) Actual photograph and (b) schematic illustration. (From ref. [2].)

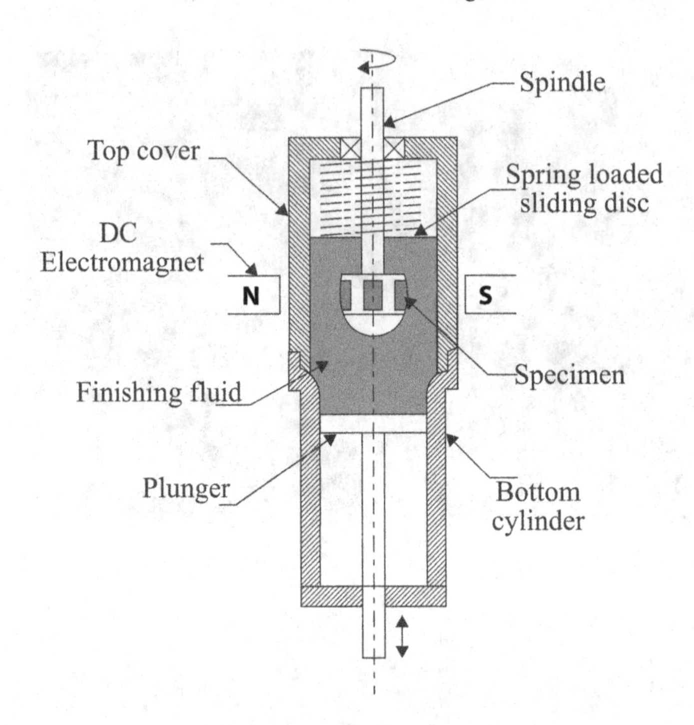

FIGURE 5.17 Schematic of magnetorheological abrasive honing process. (From ref. [7].)

The MRJF process involves the least amount of movement in all magnetic field assisted finishing processes; in this, the workpiece is attached to the spindle and can rotate on its own axis and does not involve any other degree of motion.

The third variant of MRF in consideration in this book is BEMRF, which has been developed for proper automation control of the processes. The BEMRF process and automation of the same will be discussed in detail in Section 5.7.

5.3.3.3 Motion Configuration of MRAFF and R-MRAFF

Figures 5.18 and 5.19 represent the motion configuration of MRAFF and R-MRAFF finishing processes, respectively.

The MRAFF process involves to-and-fro motion of two hydraulic cylinders, which push and pull the magnetorheological polishing (MRP) fluid over the workpiece when the fluid is under the influence of an energized electromagnet placed outside the polishing chamber. In addition to this, the R-MRAFF process involves the rotation motion of the electromagnet, which is done using a motor to impart this extra motion.

5.4 CONTROL HARDWARE AND CONTROL PANEL

To consider proper automation of magnetic field assisted finishing processes an important aspect of process automation is the use of control hardware and the control panel. The control hardware are the components that enable the user to automatically

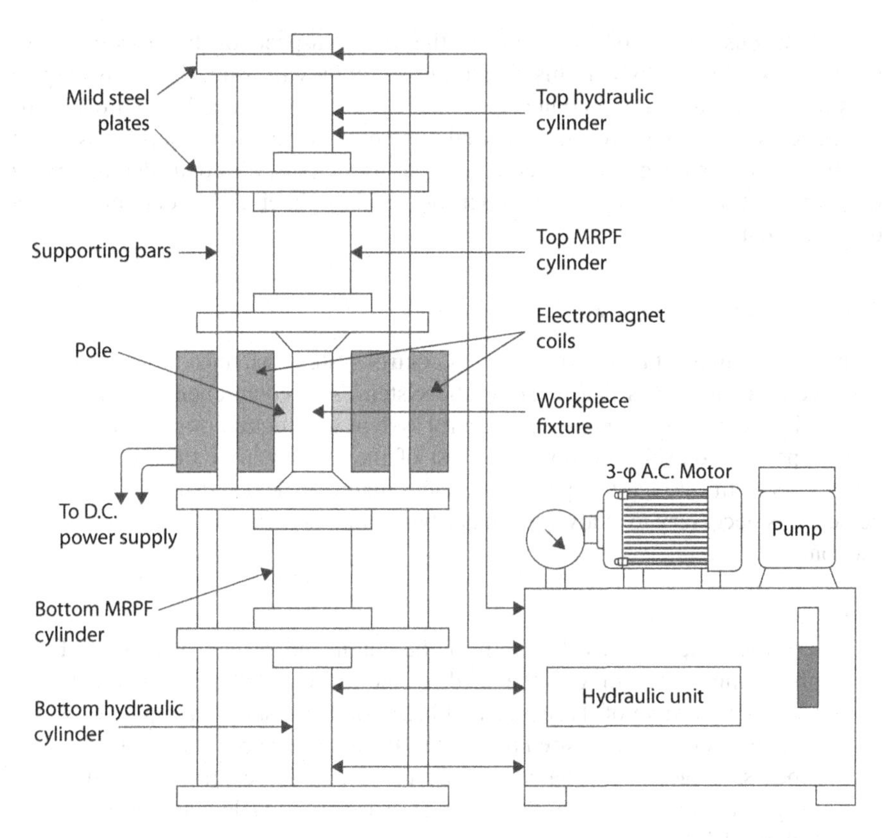

FIGURE 5.18 Schematic of magnetorheological abrasive flow finishing process. (From ref. [8].)

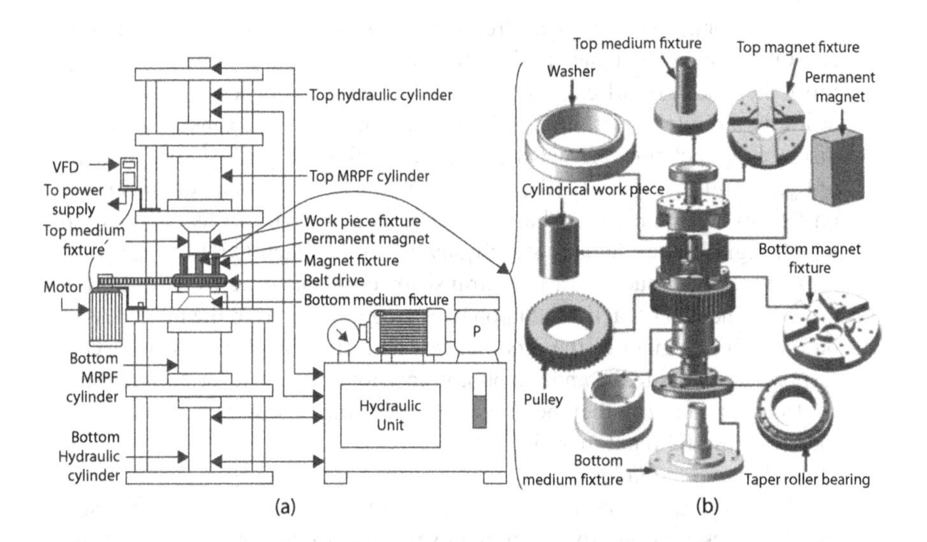

FIGURE 5.19 (a) Schematic of rotational-magnetorheological abrasive flow finishing process and (b) magnified view of the finishing zone consisting of workpiece, workpiece fixture, and magnet fixture. (From ref. [9].)

control various aspects of a process. As the motion aspects of the processes have been discussed previously in this chapter, this section will primarily focus on process parameter control and components employed for the same. There are various components in a system that are essential for proper automation of a process, these can be digital or analog devices acting as inputs or outputs, the controller and input/output modules thereof, and various sensors used to either gather data or detect a discrete event.

5.4.1 Control Hardware

To have a full understanding of a process control system, we must be familiar with the specific components that make up the system. Full comprehensions of a magnetic field assisted finishing process control system will require this familiarization of the specific control hardware. The rest of the chapter after this section deals extensively with automation options of various magnetic field assisted finishing processes. It is necessary to study control hardware and their fundamentals first in this section.

- *The controller*
  The controller acts as the brain of the automated system, which can be programmed according to the needs of the process. Once programmed, it monitors the state of the inputs and based on these states manipulates the outputs to obtain the desired results for the proper control of the automated process. Typically, in manufacturing processes, the controllers used are programming logic controllers (PLCs) or programmable automation controllers (PACs).
- *Digital input modules*
  A digital condition to be measured is one having only two states, i.e., ON and OFF. Detent switches, push-button switches, limit switches, and proximity switches are all examples of digital sensors. For a controller to be aware of a digital event, it must receive a signal from the sensor through a discrete input channel. Every digital input module, in general, has an array of light-emitting diodes (LEDs) corresponding to the connection pins it has on its physical interface. These interface pins are known as channels to which digital input devices are physically wired. Each LED has a photosensitive device, such as a phototransistor corresponding to it inside the module. The LED is activated upon receiving a signal from the input device on the input channel of the module; this activates the phototransistor corresponding to the LED, and the phototransistor provides information to the controller, which can then be processed accordingly. Figure 5.20 shows an industrial digital input module.
- *Digital output modules*
  While the digital input modules receive information from devices measuring and tracking digitally occurring events, the digital output modules do the opposite, i.e., they make the digital events occur. The information flows in the reverse order compared to the digital input devices. The controller

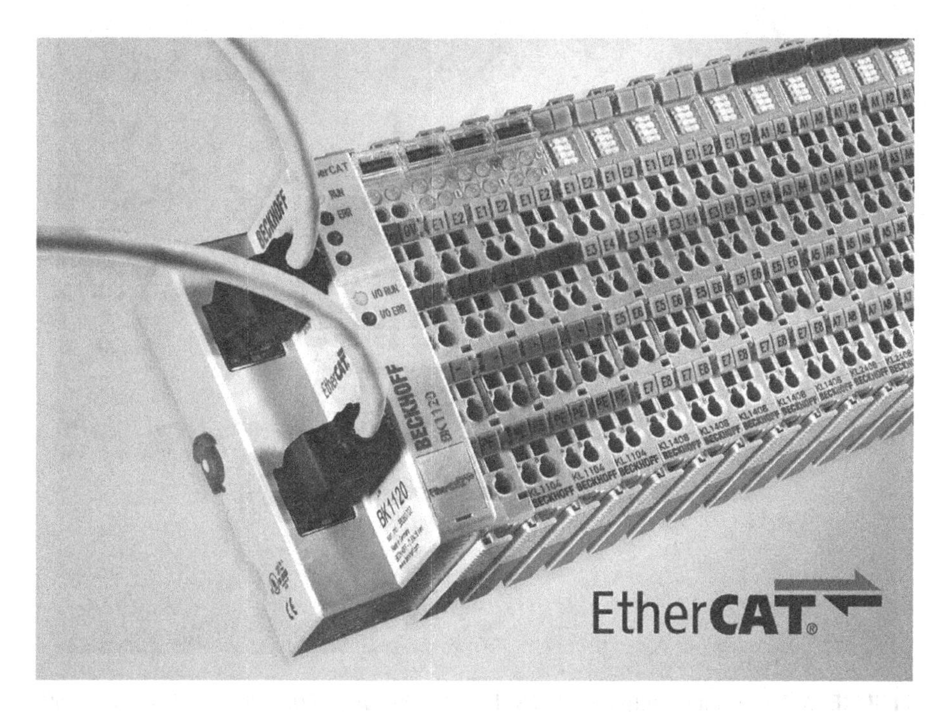

FIGURE 5.20 Digital input modules KL 1104 on BK1120 coupler. (Pic. credits: Beckhoff Automation GmbH & Co. KG.)

commands an LED to switch on and shine light on a phototransistor, which completes the circuit sending power to the digital output module channel. The channel is physically wired to devices, such as an industrial light beacon, indicator lights, buzzer, or most commonly the electromechanical relays, which closes high-power circuits to control devices, such as a motor, fan, pump, valves, etc. Figure 5.21 shows an industrial digital output module.

- *Control relay/contactor*

  A control relay, typically known as a relay, is an electromagnetic switch, which allows electrical current to flow through two conducting terminals by closing them magnetically.

  A conducting coil is energized or de-energized to make the mechanical contacts attract to close the circuit or separate to open the circuit or vice versa depending on normally open (NO) or normally closed (NC) configurations. It also protects the circuit current. Figure 5.22 shows the working of a control relay schematically.

  A contactor is also a kind of relay. The difference between a generic relay and a contactor is that the contactor can process large amounts of current, usually more than 25A. The contactors are utilized to power on or cut the power off from heavy current power drives, such as servo drives of a five-axis CNC system. The contactor functions exactly the same as the generic relay module.

FIGURE 5.21 Digital output module EL2809 on EK1100. (Pic. credits: Beckhoff Automation GmbH & Co. KG.)

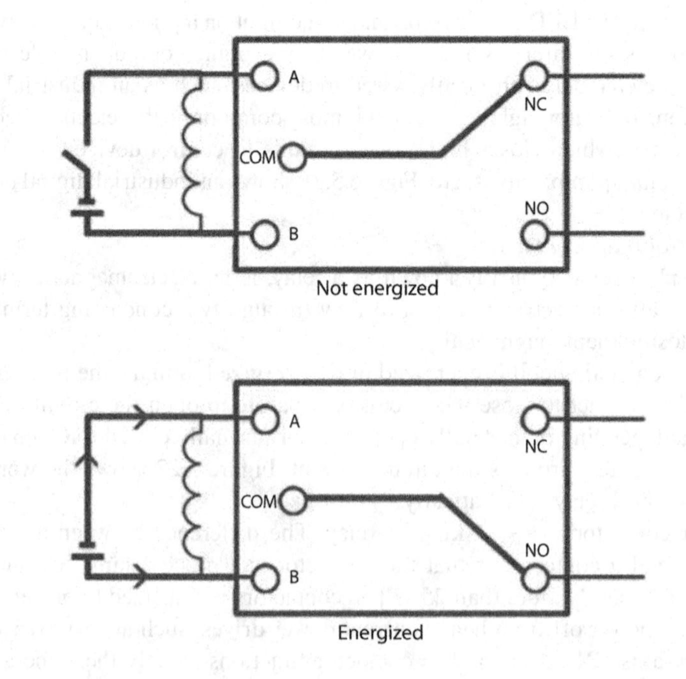

FIGURE 5.22 Schematic representation of a relay module.

- *Analog input modules*
  As described, in case of digital events the requirement is to observe the occurrence of an event and based on that provides an ON or OFF signal. However, measurement of conditions that vary continuously requires a continuous screening of the phenomenon and report the same back to the controller. Such conditions are temperature variation, pressure variation, humidity variation, etc. The sensors employed to record changes in these conditions are analog in nature and can record and send signals based on continuous variating in the physical phenomenon. The analog sensing device, such as a temperature sensor, sends continuously varying voltage/current (typically 0– 10 V or 4–20 mA) signals corresponding to the temperature variation in the environment of the process being monitored. These sensors are wired to the analog input module channel, which receives the signals and sends them to the controller. The controller can then display the calibrated values and/or act accordingly.
- *Analog output modules*
  The analog output modules provide continuously varying voltage/current signals at the output channel of the module, which are wired to devices that can be controlled via analog programming. Such devices adjust their operation corresponding to the voltage/current signals received from the analog output module channel. Devices such as variable speed drives that control motor speed usually are programmed using analog output signals from the controller.

  Pump speed/frequency control (MRP fluid supply), automated magnetizing current variation (for electromagnets), DC motor (e.g., in stirring action for MRP fluids), speed control, automated coolant temperature control, etc. are some examples of applications that require analog output signals from the controller. Figure 5.23 shows images of industrial analog input and output modules.
- *Power supply*
  A power supply becomes an integral part of any automated system, and this is because the controller output signals are for information only and are not strong enough to run a motor or pump, like the servo and stepper drives, which are a power source for the motors they control. The power supply provides power to devices that require higher current, which the controller is unable to provide. The examples include powering up a high-power–rated direction control valve that moves the hydraulic cylinder of, say, the MRAH process.

  The digital signal from the controller will not be able to supply enough current to the valve solenoid to switch it on. Thus, to switch the valve solenoid on the controller will provide the digital signal via the digital output module, and this signal will switch on a relay, and that will then switch on a high-power circuit for the valve solenoid to get energized and perform its action properly.
- *Push buttons/selector switches/indicator lights/pilot lights*
  The actions performed by an automation controller are a result of a set of logic programming, which monitors the state of the input channels and,

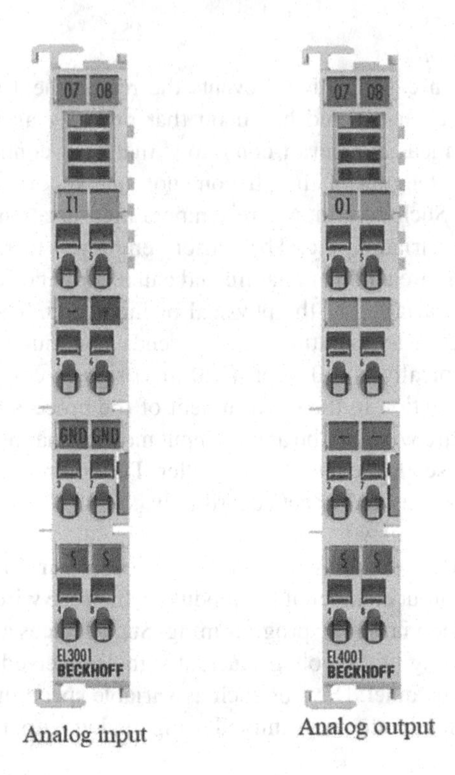

FIGURE 5.23 Analog input module (EL3001) and analog output module (EL4001). (Pic. credits: Beckhoff Automation GmbH & Co. KG.)

based on that, manipulates the output channels available on a system. The input actions are either an output from a sensor, such as a proximity sensor, limit switch, etc., or an action performed by an operator. The actions from an operator are given to the controller by means of buttons or switches that are manipulated manually; these are of different types and are covered very briefly here.

The *push buttons* are momentary input devices and provide an input signal to the controller for as long as they are pressed by the operator. The *selector switches* are detent-type switches, which remain in their state unless changed by the operator, i.e., once switched ON from OFF, the switch will remain ON unless the operator acts to willingly switch it back OFF. Figure 5.24 shows an image of industrial push buttons and selector switches.

Like digital inputs, there digital outputs that are required within the system to make the system interact well with the operators. Other than the digital outputs required for controlling the system, there are indications required to make the user aware of the state of the machine. These indications are typically made using LED *indicator lights* or *pilot lights*.

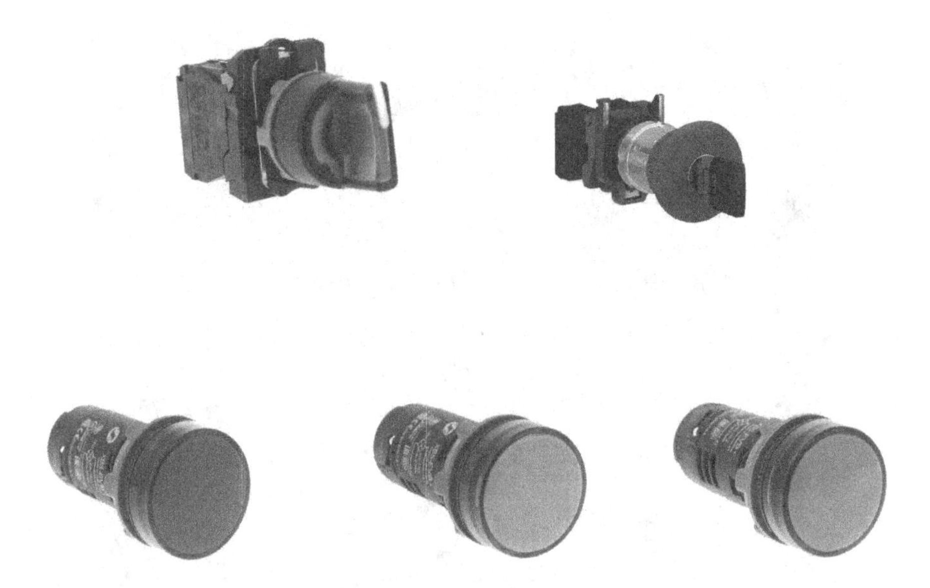

FIGURE 5.24 Industrial control switch, emergency switch, and indicator lights. (Pic credit: Schneider Electric.)

5.4.2 Control Panel*

A *control panel* is an assembled standard housing of the various control hardware components such as motor (servo or stepper) drives, controller (PLC) power supply, control relays, MCBs, and related control devices including pushbutton stations, selector switches, timers, switches, with associated wiring, terminal blocks, pilot lights, and similar components.

The terminals and terminal blocks therein are important for wiring of each of the components together. They are color-coded, and terminal block diagrams are made with terminals numbered so that wire tracking is easy, and maintenance of the system is hassle-free. There are components that use AC power, and there are those that run on DC power inside a control panel. The control panel design is kept as such that the devices with different type of power are separate and wires running through them also have separate routes to avoid power hindrance.

There are various design considerations considered to design a proper industrial control panel. Some of the design considerations that are important to consider are wire sizing and component types, overcurrent protection, labeling, and terminal numbering. An important aspect of wiring a control panel is grounding, which ensures the panel is properly connected to the ground electrically. Electrical grounding is a backup pathway that provides an alternating route for the current to flow back to the ground if there is a fault in the wiring system. It facilitates a physical connection between the ground and the electrical equipment and appliances in the control panel. Figure 5.25 shows an image of a typical control panel.

*The discussions around control panel in this chapter are purely for the purpose of general information only and **must not** be taken as a guidance for a full-fledged

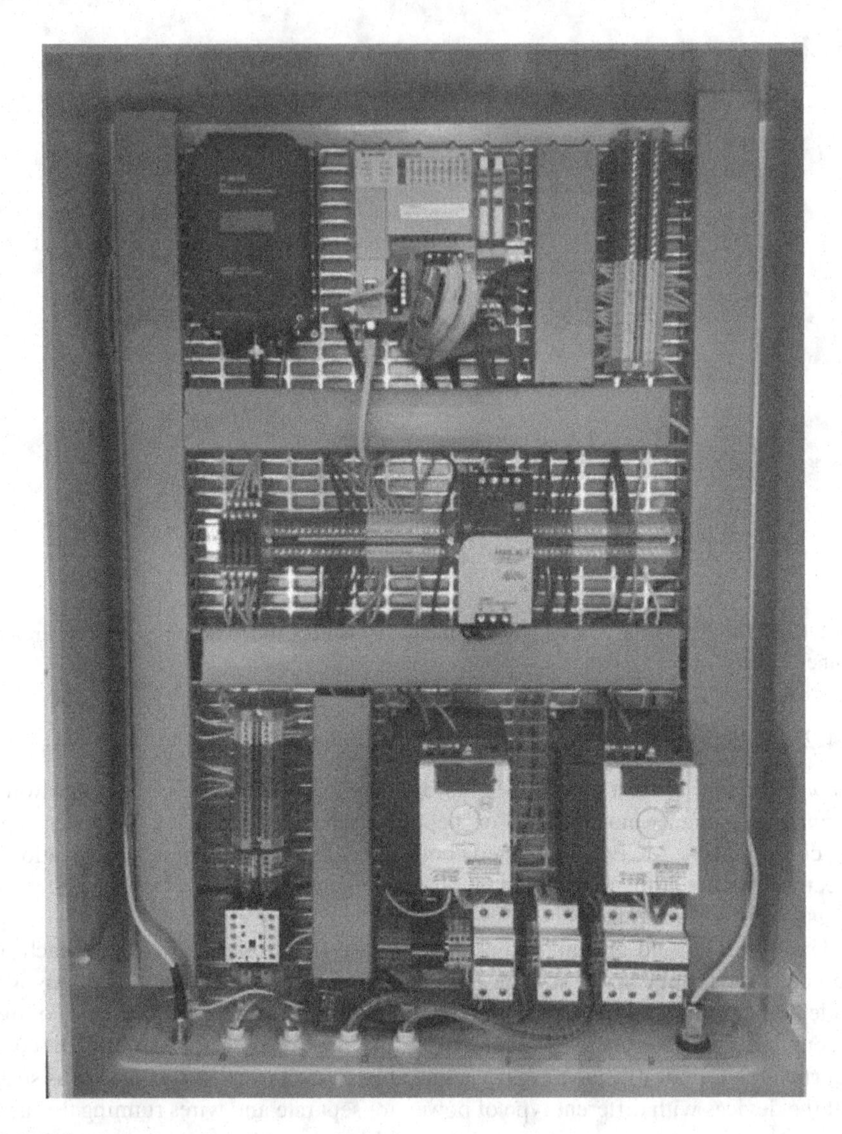

FIGURE 5.25 An industrial control panel.

control panel design. To properly design and use a control panel, an expert advisor must be consulted with.

5.5 USER INTERFACE AND PROGRAMMING

While automating any mechanical system to a good extent a user interface is a must. User interfaces are designed to facilitate the interaction of the automated system with the person using it. They should be designed to make the automation of the system easier. This way, the users can operate the automated system, are aware of what happens when the system is in operation, and are happy to control the system at their own convenience. For magnetic field assisted finishing processes, certain minimum functionalities should be included to control the process parameters as well as the motion parameters in the automation of these processes. The minimum functionalities include the digital user controls, analog user controls, displays for analog and digital events being sensed, and motion position of each movement involved. A typical user interface operates in two modes, viz. automatic mode and manual mode.

5.5.1 Automatic Mode

The automatic mode of the user interface is for setting up all the parameters of the system before the process starts, and then just by pressing the start button the whole process runs automatically. In magnetic field assisted finishing processes this includes setting up of the desired parameters, such as magnetizing current, spindle RPM, working gap, etc., before pressing the start button. Motion parameters, if any, are also selected prior to start. It involves different ways of doing it: either a code script is written for the whole duration of the process and all parameter values are included at desired time intervals according to the code, or it includes a repeatable code script running repeatedly in cases of the process having actions that are repetitive in nature. The parameters can also be utilized using variables, which are updated in real time when the process is running.

5.5.2 Manual Mode

As the name suggests the manual mode lets the user control the system manually. This means that each individual component of the system is available to the user to be manipulated. Manually means that the manual mode user interface has all the entities for the system to be handled by individual buttons on the manual mode interface. The *automatic mode* and *manual mode* interfaces will be described in detail in a case study later in this chapter.

The user interface/s, however, are only for the user interaction and ease of use for the person using the automated system. There is a great deal of effort required to communicate the automation control hardware, e.g., a PLC. It is the communication between a human programmer and the PLC that enables the PLC inputs and outputs to behave in the manner they should be behaving to automate a mechanical process. There are various PLC systems available in the market, and each of them is programmed using their own software. However, the standard for writing the PLC programs remains the same and is updated regularly to include advances in technologies in the rapid advancement era. Typically, the latest standard followed worldwide for the programming of PLCs is IEC 61131.

This chapter does not intend to deviate much from the topic of automating the magnetic field assisted finishing processes by going into the depths of programming standards and languages. A good guide for PLC programming standard IEC

61131 and related text can be found in literature archives, such as ref. [10], to find out answers to the following questions for the users:

- How do you program in accordance with IEC 61131? What are the essential ideas of the standard, and how can they be applied in practice?
- What are the advantages of the new international standard IEC 61131 compared with other microcontroller programming or PC programming?
- What features must contemporary programming systems have to be consistent with IEC 61131 and to fulfil this standard?
- What do users need to look for when selecting a PLC programming system?
- What criteria are decisive for the performance of programming systems?

5.6 FEEDBACK SYSTEMS

Among the automated systems, the most advanced systems are those that are controlled in a closed-loop manner. The closed-loop control enables systems to achieve higher accuracy and better repeatability. The most important component of a closed-loop automated system is a feedback of the output generated. The feedback means monitoring the generated output and providing it back to the controller to take necessary corrective action. The quality of magnetic field assisted finishing processes is quantified based on surface roughness achieved. Thus, a good automated magnetic field assisted finishing process must have a feedback system in it. The following subsections give a brief overview of feedback systems and related information thereof for magnetic field assisted finishing processes.

5.6.1 EQUIPMENT

The surface roughness monitoring equipment primarily used for this purpose is quite diverse, both in terms of size as well as the working principle. Over the years many techniques have been employed for this purpose, but with time, technology has taken over a few of the traditional techniques owing to their bulky size and slow response thereof. Launhardt et al. [11], in a recent study, described the state-of-the-art systems that are used for measurement of surface roughness. They classified these systems into four categories: tactile profile measurement, focus variation measurement method, fringe projection technique, confocal laser-scanning microscope.

Tactile profile measurement is the most traditional and most reliable of the lot and uses a stylus to physically scan a surface. The stylus disturbance due to surface asperities is recorded and plotted to find out the surface roughness being measured. In the focus variation method is an axial scanning measuring technique that is both optical and area-based. The fringe projection technique is based upon a light pattern of equidistant, parallel fringes, projected onto a surface. In a confocal microscopy, a collimated white light source beam is guided by a dichroitic mirror onto a microscope objective. This focuses the beam onto the focal plane, where a specimen is sited. The reflected beam is collected by the microscope objective, and color-sensitive detectors capable of detecting colors at very close wavelength tells the distance between two successive values.

All the other methods are outdated when it comes to fast, reliable data and features that enable online measurement of the surface being able to provide data to a controller. In this regard, the four techniques mentioned previously provide the data, but the bulky size of the first three makes them difficult to be used in conjunction with the magnetic field assisted finishing processes. The fourth method, i.e., confocal microscopy, has been successful in being transformed into compact lightweight systems easy to be integrated with manufacturing systems.

5.6.2 Data Acquisition and Analysis

For automating the magnetic field assisted finishing processes in a closed-loop manner, surface roughness data must be acquired. As described in the previous section, many different types of equipment can be used to achieve this. However, to have a system that is automated and works well in integration with all other parts of the magnetic field assisted finishing process setup, a compact system that can capture data and transferring it to the controller is required. The confocal microscopy-based chromatic confocal sensor is very compact, lightweight, and can easily be integrated with any setup both mechanically, as well as software-wise.

The confocal sensor is constructed in such a way that it takes white light from a source and breaks it into various colored lights using an aberration of lenses. The multicolored light then is impinged on the surface being monitored. The reflected light is collected back in the confocal sensor by color-sensitive receptors, which are capable of distinguishing light-color resolution of 10 nm wavelength. This way, the peaks and valleys on a surface with a resolution of 10 nm are recorded and sent to the confocal controller first to run algorithms to remove any noise data and filter any undesired recording. The filtered data is then transferred to the main process controller or a PC-based controller in the form of. CSV files, which can be read by software programs controlling the main magnetic field assisted finishing process.

The data thus acquired is then analyzed in light of the current process and the required target of roughness against the initial roughness value, and critical actions can be taken by the controller to decide what to do further in the process. This gives a chance to the process controller to adjust the process parameters according to the current state of the workpiece without dismounting the workpiece from its fixture. The data acquisition in such processes and the automation thereof can only be in situ because of the MRP fluid that is involved in finishing the surface. The fluid becomes a barrier to capture data as it covers the surface, and a workpiece cleaning system must also be there to clean the surface of MRP fluid before data acquisition process can take place.

5.6.3 Control Action

The acquired data after being analyzed can be used in a variety of different ways depending on the user requirement. The process is programmed as per the requirement, and based on the various conditions set in the program a control action is taken to ensure proper automatic functioning of the process. As an example, the process may be required to stop at a point where the roughness has been reduced to half of the initial value or a quarter. All this is set at the very beginning when the controller is being programmed to carry out the process automation. Once the feedback is received the controller checks its library of commands to find out the next available action to be performed further. It could be a control action of finishing further for another 2 hours or stop the finishing action and tag it complete.

5.7 AUTOMATION OF BEMRF PROCESS: A CASE STUDY

The description of all the control hardware and the software thereof previously discussed in this chapter has focused mainly on individual functions of all of them and how each of those can be used to perform one or more tasks required to automate a magnetic field assisted finishing process. The description thus far has been quite generic, and their role in automating a magnetic field assisted finishing process can still be further explained to make the reader understand it in more detail.

Automating a magnetic field assisted finishing process can be a tedious task, and if not done right can lead to undesired problems halting the essential work for long durations. A good knowledge of what control hardware to select for automating a particular process, integration of those components with the mechanical setup of the process, choosing a desired controller and programming the automation sequence all involve key decision-making and require thorough knowledge of both sides of the system. To take the audience of this book a step closer to being able to automate the magnetic field assisted finishing processes, a case study involving the automation of BEMRF process is offered here. Using the case study, the readers will be taken to all the steps required to automate a magnetic field assisted finishing process.

With the BEMRF process having been already discussed in previous chapters and with detailed process explanations available in archival literature [12–26], we will directly dive into the structure of automation in the BEMRF process [15]. Like all automatically performing processes this one also includes the same components, i.e., it has a physical setup, a software to control that, and a controller that takes actions to make the whole thing run smoothly.

5.7.1 Physical Setup of BEMRF

The physical setup of the five-axis CNC BEMRF machine has three linear and two rotary positioners, which provide three linear movements along X, Y, and Z axes and two rotary movements along B and C axes, respectively. The two rotation axes B and C are orthogonal to each other. B-axis is parallel to Y-axis, and C-axis is parallel to Z-axis, while the three linear axes X, Y, and Z are orthogonal to each other.

The BEMRF tool is an electromagnet kept inside a closed cylinder filled with transformer oil, which acts as a coolant; it has a spindle passing through it which is coupled with the spindle motor at the top through a pulley and belt arrangement and has a tip at the bottom. The spindle has a hollow passage through which MRP

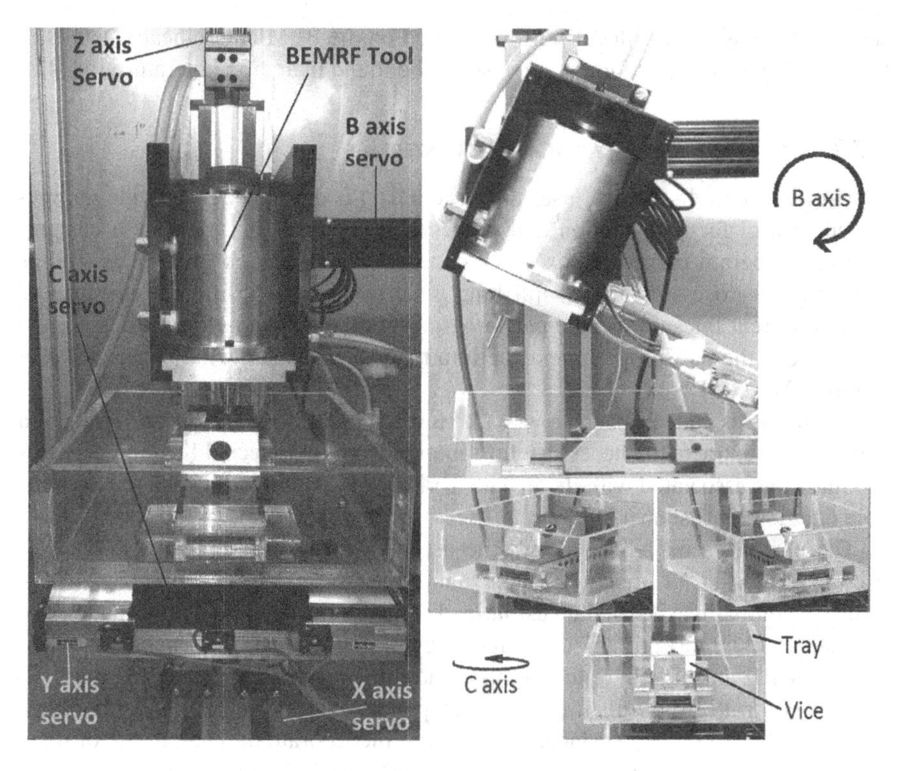

FIGURE 5.26 Ball end magnetorheological finishing (BEMRF) setup: arrangement of five motion axes and the BEMRF toolhead. (From ref. [15].)

fluid is supplied to the tip. The complete arrangement of the five axes along with the BEMRF tool is shown in Figure 5.26.

The five axes and the spindle are coupled with servomotors, each of which is connected to an Ethernet power link (EPL) servo drive. All the drives are controlled by a PAC motion controller connected with a personal computer (PC).

The CNC BEMRF system requires a six-axis motion controller, servo drives and motors, digital input/output modules, and analog input/output modules. The six drives and motors, required for motion control, include five axes (X, Y, Z, B, and C) and one tool spindle. Six digital inputs are required to read the status of the end-limit sensors of the three linear positioners (X, Y, and Z).

The end limits are sensed by magnetic proximity sensors placed at both ends of the linear positioners. Three digital outputs are required to enable/disable the electromagnet power supply, stirrer motor, and the VFD, which controls the speed of the fluid delivery pump. One analog input is required to read the output of the resistance temperature detector (Pt100) placed inside the electromagnet coil. To control the stirrer speed, magnetizing current, and the speed of the fluid pump, three analog outputs are required.

The physical setup also consists of workpiece cleaning system details, which can be found from archival literature in [27]. The roughness measurement in the BEMRF setup is done using a confocal sensor mounted on the BEMRF toolhead. Details of the full roughness measurement process can be found in archival literature in [14].

5.7.2 Software and Graphical User Interface

A Visual C#-based graphical user interface (GUI) is developed to provide an integrated control for both motion and process parameters. The GUI has a customized part-program parser that executes standard CNC codes for motion as well as facilitates in-process control of finishing force by providing different current values in different lines of a single part-program. As the upload button is pressed, the parser loads the part-program text file into the parser display window as shown in Figure 5.27.

The parser is coded in such a way that it reads one line of the part-program at a time. It then extracts the necessary information from the selected line and executes the functions associated with them. The working of the parser is explained with the help of a flowchart shown in Figure 5.28.

The part-program parser allows the user to work with the same CNC codes used in standard CNC machines. This makes the five-axis CNC BEMRF machine industry compatible. Also, the customized controller developed here has an added advantage in the form of "E" code (for magnetizing current) that enables the five-axis BEMRF machine here to automate the finishing of 3D surfaces through in-process control of the finishing forces. Depending on the surface and roughness profile of the workpiece surface, the user can incorporate the magnitude of magnetizing current (finishing forces) in the standard CNC codes and execute them using the customized controller to achieve complete automation of the finishing process.

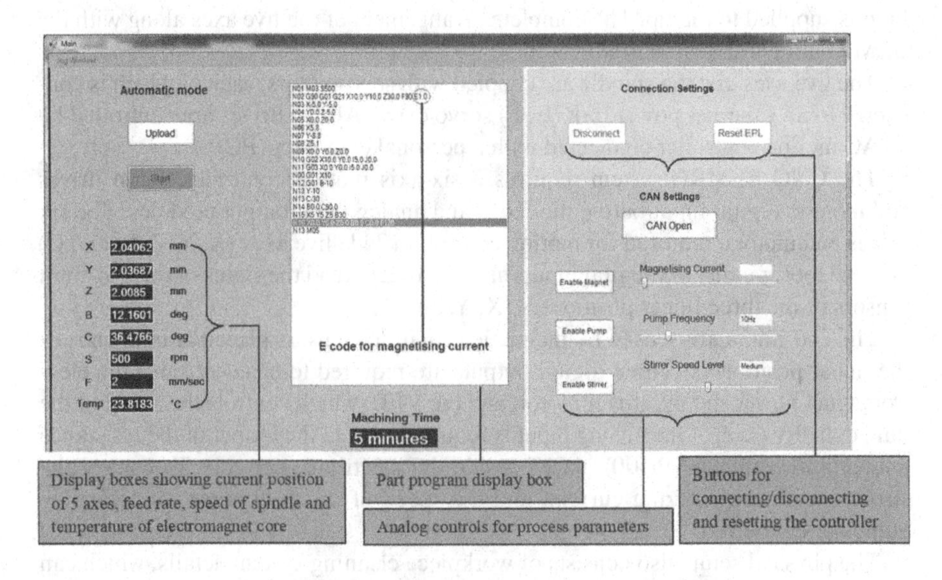

FIGURE 5.27 User interface for BEMRF process. (From ref. [15].)

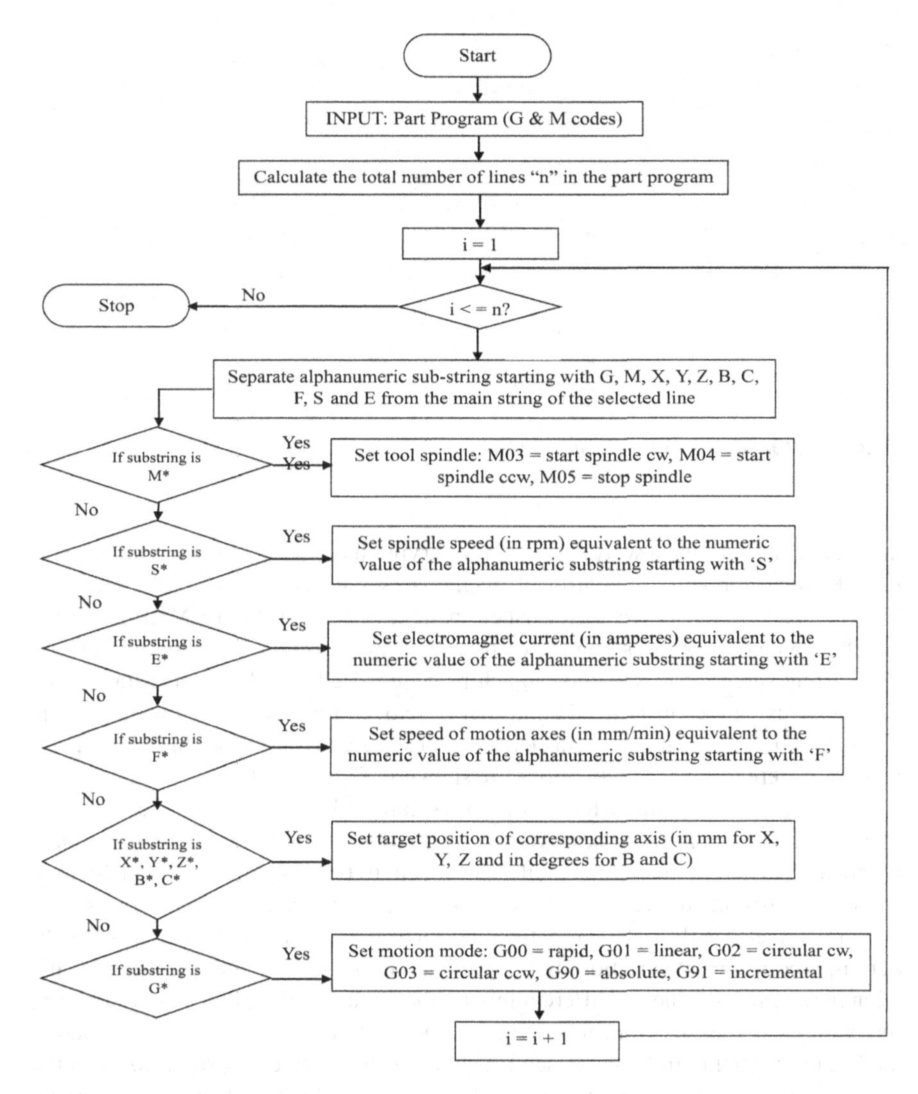

FIGURE 5.28 Flowchart for part-program parser. (From ref. [15].)

5.7.3 Controller and Finishing Results

The five-axis CNC BEMRF machine controller customized for automated finishing here not only maintains a uniform working gap in the finishing region but also has the flexibility to vary the magnetizing current over different surface profiles to realize enhanced finishing results. To study the effect of working gap and current on surface finish, three mild steel test specimens were finished separately on different machines. The first sample was finished on a three-axis CNC BEMRF setup. A constant value of magnetizing current was used to finish the entire surface profile. The second specimen was finished on a five-axis CNC BEMRF machine that helped

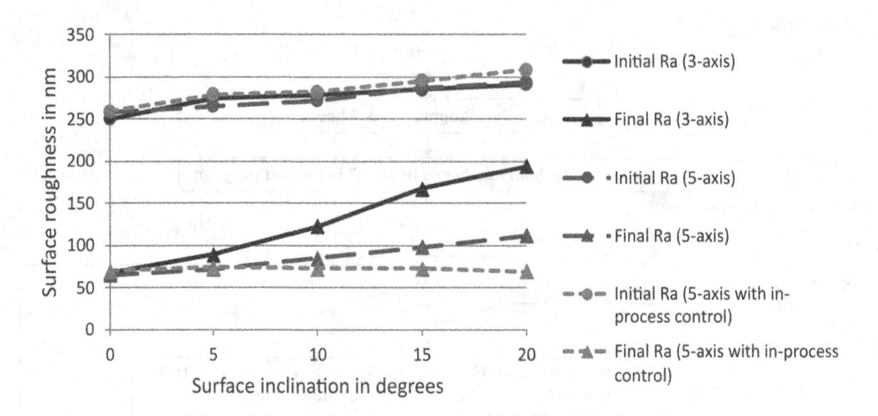

FIGURE 5.29 Comparison of surface roughness. (From ref. [15]).

maintain a near-uniform working gap over varying inclination of the workpiece surface. However, in-process control of magnetizing current was not employed here. Finally, the third specimen was finished using a five-axis CNC BEMRF machine, which provided in-process control of both working gap and magnetizing current. The current values on this five-axis machine were varied over different surface inclinations by using the customized controller developed for automated finishing of 3D surfaces. The initial and final surface roughness value (Ra) of the three test specimens at different surface inclinations are shown in Figure 5.29.

From the results obtained, for a three-axis setup, as the inclination angle increases, final roughness value increases. Also, the variation of the final surface roughness obtained at different inclination angles is very high (68–194 nm). As the inclination angle increases, the unevenness of the working gap below the tooltip increases.

For the five-axis BEMRF setup, the final surface roughness values obtained without in-process current variation show significant improvement, but still there is variation in roughness values at different inclination angles. Conceptually, it is supposed that with even with the working gap maintained below the tooltip, at all inclination angles, the variation in final surface roughness value should be proportional to the initial roughness value. However, the difference in roughness values is considerably higher. This may be because of the reduction in normal finishing force due to the gravitational force acting on the MRP fluid over inclined surfaces.

This reduction in normal force can be compensated for by an increase in magnetizing current proportional to the surface inclination angle of the workpiece. Also, higher value of forces is required to obtain uniform finish when the initial surface roughness is higher as is the case here. This is precisely what has been achieved with the mild steel test sample being finished with the automated five-axis CNC BEMRF with in-process control of magnetizing current.

As seen from roughness comparison in Figure 5.29, almost equal surface finish is obtained over the entire surface profile irrespective of the value of the initial surface roughness or the inclination angle. This is due to the uniformly maintained working gap and a controlled magnetizing current over the entire surface profile achieved by

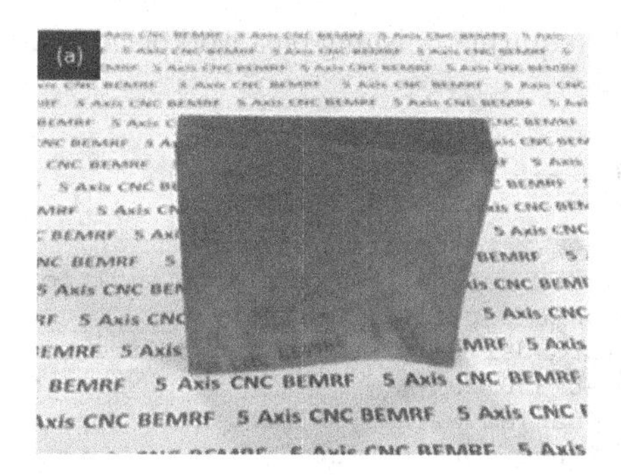

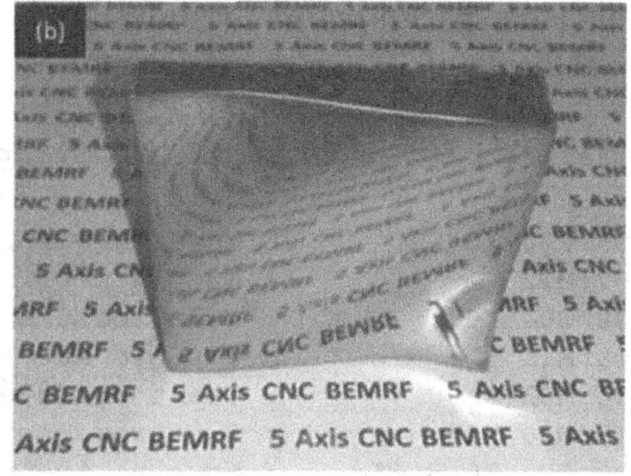

FIGURE 5.30 Photographs of (a) initial surface and (b) final surface finished by automated five-axis CNC BEMRF machine with in-process control of magnetizing current. (From ref. [15].)

the automated 3D finishing using the five-axis CNC controller customized for the BEMRF machine.

Using standard CNC codes for a 3D part, the controller facilitates change of finishing forces for different areas of the workpiece surface profile through "E" code. The actual photographs of the initial and final surface finished with automated five-axis CNC BEMRF machine with in-process current control are shown in Figure 5.30, where a uniformly finished 3D freeform surface is obtained as compared to the initial surface.

5.7.4 Part-Program–Based Control of the BEMRF Process

The previous sections established the capabilities of the BEMRF process as an automatically controlled process and provided details of how the automation brings out better surface finishing results compared to the process when controlled through semi- or no automation. The control software is developed so that the functioning is fully controllable using part-programs. The functioning of the developed five-axis system for BEMRF process is divided into three cycles: finishing cycle, workpiece cleaning cycle, and measurement cycle, all by a single part-program [28]. The cycles along with their respective part-programs are as follows.

5.7.4.1 The Surface Finishing Cycle and Associated Part-Program

The surface finishing cycle governs the control of motion control hardware for the five-axis arrangement, the BEMRF toolhead for magnetizing current control, the MRP fluid delivery system, and the refrigerated water bath for coolant to the electromagnet. The finishing cycle being the main central operation of the BEMRF process consists of the motion control of all five axes, spindle control using conventional CNC codes. Apart from the conventional operations the in-process control of the magnetizing current has also been custom developed specifically for BEMRF process. The finishing cycle thus developed will govern primarily the motion part which relates to the geometry of the workpiece, the basic process controls, and the magnetizing current controls for the electromagnet. Following is an example of a standalone part-program for finishing cycle requiring finishing on a workpiece for a fixed length of 100 units:

```
%Example part-program for finishing cycle

N05 G00 X0 Y0          ; Move to workpiece zero position.
N07 M12                ; Enable stirrer and deliver MRP fluid to tooltip.
N08 G04 X05            ; Delay of 5 seconds for fluid delivery.
N09 M13                ; Disable MRP fluid delivery system and stirrer.
N10 M03 S300           ; Start spindle at 300 RPM.
N15 M10 H45            ; Electromagnet enabled with 4.5A current.
N20 G01 Z0.8 F10       ; Move to the defined working gap.
N25 G90 G01 X100       ; Move x-axis to 100 units.
N30 X0                 ; Move x-axis back to 0.
N35 M05 M11 H0         ; Spindle stop, electromagnet disable, current 0.
N40 G01 Z35            ; Move tool upwards after finishing.
N45 M30                ; program end.

*The above program is a simple example of finishing cycle and not a
full-fledged finishing CNC code.
```

5.7.4.2 Workpiece Cleaning Cycle

In the automation of the BEMRF process, the need to remove the workpiece every time the finishing cycle is completed should be eliminated and the measurement of roughness parameters must be carried out; this ensures that the complete finishing process is carried out till the desired finish is achieved with the workpiece being fixed in its place throughout the process. To carry out measurement of surface finish within the BEMRF setup, the workpiece must be cleaned automatically. Manual removal of MRP fluid cannot be done in an automatic system.

To achieve this, a workpiece cleaning system was developed, which had certain controls that can also be controlled through part-programming [27]. The workpiece cleaning cycle requires movements of the enclosure box, and the jet nozzles along with the air and kerosene jet pump controls. Motion of the worktable, which moves the workpiece along with it, is also required at the beginning and end of the cleaning cycle. As soon as the workpiece cleaning cycle is called, the worktable, after ensuring safety from BEMRF tool, is moved to the cleaning station. At the cleaning station, the enclosure box moves to enclose the workpiece, the kerosene jet is started, followed by to-and-fro motion of the nozzle for a fixed period. After cleaning from kerosene, the air jet is started, and the nozzle again moves to and fro to clean the residual kerosene droplets.

The workpiece is moved away from cleaning station after completing the cleaning process. Automated cleaning process using part-programming is achieved by specially associating the movements of enclosure box, jet nozzles, jet pump, and air pressure controls with M codes. Custom M codes are defined for each action required in the cleaning process; these are specific to workpiece cleaning systems. An example CNC code for the workpiece cleaning cycle is shown as follows. In it, only a single stroke of nozzle each for kerosene jet as well as the air jet is shown for indicative purposes; however, in actual practice there are multiple strokes, and the CNC code is written accordingly.

```
%Example part-program for workpiece cleaning cycle

N05 G90 G01 X70 Y-70 F100        ; Move worktable to cleaning station.
N10 M15                          ; Enclose workpiece.
N15 M17 M21                      ; Start kerosene jet, nozzle forward.
N20 G04 X5                       ; Delay
N25 M18                          ; nozzle backward
N30 M22                          ; Stop kerosene jet.
N35 M17 M26                      ; Start air jet, nozzle forward.
N40 G04 X5                       ; Delay
N45 M18                          ; nozzle backward
N50 M27                          ; Stop air jet.
N55 X0 Y0                        ; Move worktable to finishing station.
N60 M30                          ; Program end
```

5.7.4.3 Roughness Measurement Cycle

The measurement of surface roughness in the BEMRF process can only be conducted in situ. For in situ surface roughness measurement, the chromatic confocal sensor, lightweight and compact in size, is used with the BEMRF toolhead. The controls for roughness measurement cycle are also mapped with CNC part-program codes for better control.

The roughness measurement cycle requires movement of the confocal axis and the ON/OFF triggering of data acquisition by confocal sensor. Once the workpiece cleaning cycle is completed, the measurement of surface roughness is required immediately. For this purpose, the roughness measurement cycle is called soon after the workpiece cleaning cycle. First, at the start of the roughness measurement cycle the worktable is moved to the roughness measurement station, i.e., the area on the workpiece under consideration for finishing and measurement is moved underneath the confocal sensor. The confocal sensor controlled by a stepper axis is moved towards the workpiece and reaches the measuring range.

Like the workpiece cleaning cycle, the roughness measurement cycle also involved definitions of new M codes for roughness measurement cycle-specific tasks such as movement of confocal axis, triggering of data acquisition etc. An example CNC code for the roughness measurement cycle is shown as follows.

```
%Example part-program for roughness measurement cycle

N05 G90 G01 X-50 Y0 F100         ; Move worktable to measuring station.
N10 M24                          ; Move sensor in range.
N15 M32                          ; Start data acquisition.
N20 G01 X-54                     ; Sampling length movement.
N25 M33                          ; Stop data acquisition.
N30 M25                          ; Move sensor up.
N35 X0 Y0                        ; Move worktable to finishing station.
N40 M30                          ; Program end.
```

5.7.4.4 Integrated Part-Program for All Three Cycles

Individual control of each cycle as explained earlier is used when any of these actions are required separately. However, the automated performance of BEMRF process requires these cycles to operate in tandem, and therefore it is required to have an integrated CNC, code which can have all these tasks performed in a single program. A program can be written that can contain all these codes in a sequence required to get the BEMRF process automated. But such a program will become too long and prone to errors. Thus, automation of BEMRF process is obtained by using certain subroutine calling functions, which are explained as follows:

The CNC code files are used with an extension. NC (read as dot NC), the cycles are called as a subroutine by their file names. The sequence call subroutine function termed as "L SEQUENCE"[1] is used in the main program from which the subroutines are called. The main program file is uploaded to the CNC program reader and the program is then started. The CNC code of the main file is executed line by line and whenever it encounters the following line:

```
L SEQUENCE [NAME="filename.nc", REPEAT = 06, N05, N25]
```

The program then jumps to the subroutine file name of which is given in the L SEQUENCE call, and the subroutine is repeated the number of times specified in the call line; also a specific section of the subroutine file can be called by specifying the serial number of the start and stop lines of subroutine file. For example, in the L SEQUENCE line mentioned above the filename.nc file will be called as a subroutine and lines N05 to N25 all will be executed six times as specified. For an automated BEMRF cycle, individual cycles of finishing, cleaning, and measurement are to be called in a similar manner. An example of the integrated CNC code for automated BEMRF cycle is given below.

```
N10 G00 X0 Y0 Z0                            ; Rapid positioning.
N20 M10 H30                                 ; Enable magnet, set current to 3A.
N40 G01 Z0.8 F10                            ; Move to machining zone.
N50 M03 S300                                ; Start spindle, set spindle speed.
N60 L SEQUENCE [NAME="Finishing.nc", REPEAT = 20, N01, N02]
                                            ; Finishing cycle.
N70 L SEQUENCE [NAME="Cleaning.nc", REPEAT = 1, N01, N12]
                                            ; Cleaning cycle.
N80 L SEQUENCE [NAME="Measurement.nc", REPEAT = 1, N01, N08]
                                            ; Measurement cycle.
N90 G01 X0 Y0 Z0 F200                       ; Return to zero.
N100 M11                                    ; Magnet disable.
N110 M30                                    ; Program end.
```

The displayed code is an example of the integrated CNC code for automated BEMRF cycle; in this, the finishing cycle named as finishing.nc will only have the movement of five axes according to the geometry of workpiece. The cleaning.nc and measurement.nc files are the same as part-program examples given for workpiece cleaning and roughness measurement cycles, respectively. Thus, the integrated CNC code executes all three cycles in tandem to achieve complete automation of the BEMRF process.

---

[1] While the control system can be any PLC / or PAC, the automation of the BEMRF process was realized using PC-based control system. Beckhoff IPC (CP6231-0001-0050) along with TwinCAT software at CNC level was used.

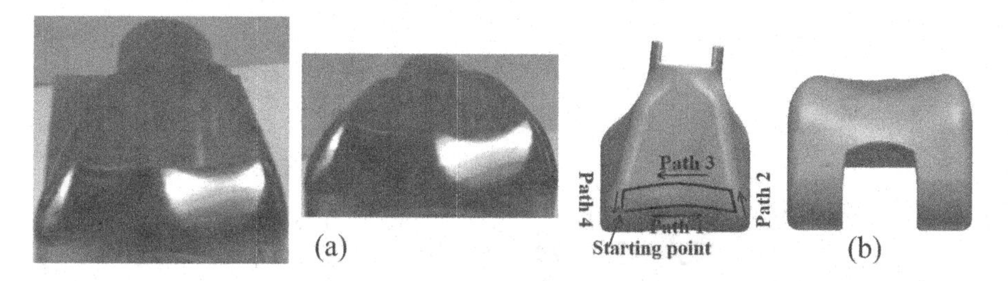

FIGURE 5.31 (a) Knee prosthetic, (b) CAD model of knee prosthetic. (From ref. [28].)

5.7.5 Testing and Results of Automation of BEMRF Process

A freeform knee prosthetic shown in Figure 5.31(a) is finished along the path shown in Figure 5.31(b) for testing of the integrated part-program. The positional coordinates of three linear axes (X-Y-Z) and one rotary axis (B) are altered to finish the work surface.

It must be noted that during the finishing of the knee prosthetic, the position of the rotary C-axis is unchanged while the X-, Y-, Z-, and -B axes vary as shown in Figure 5.32.

The position data of X-, Y-, Z- and B-axes is continuously logged and plotted at selected points along the paths 1–4, as shown in Figures 5.33–5.36, respectively.

Figure 5.37 shows actual photographs of the i5-B CNC BEMRF system at different stages of the integrated part-program.

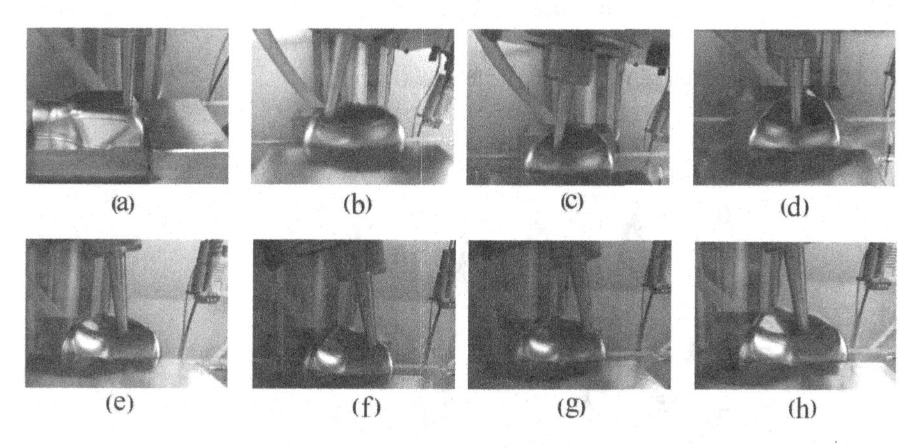

FIGURE 5.32 Different positions of the BEMRF tooltip during BEMR finishing of knee prosthetic workpiece: (a) C-axis: 90°, (b) first sampling point, (c) second sampling point, (d) workpiece zero, (e) fifth sampling point, (f) sixth sampling point – end of path 1, (g) path 2: after movement of Y-axis, and (h) eighth sampling point on path 3. (From ref. [28].)

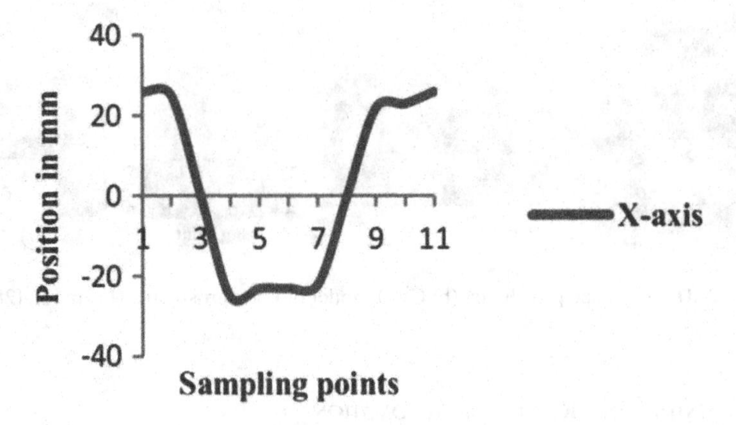

FIGURE 5.33 Position of X-axis along paths 1–4. (From ref. [28].)

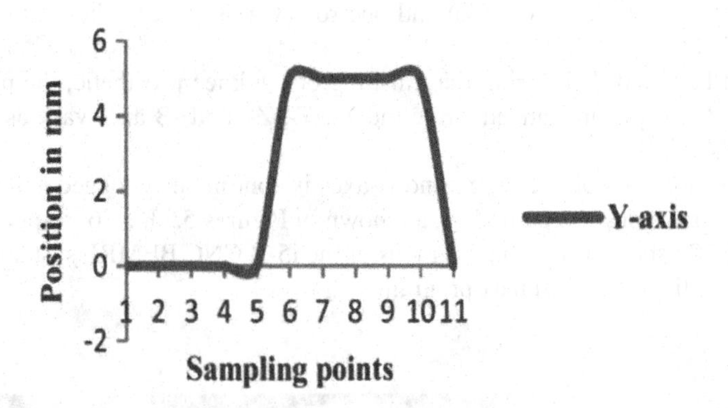

FIGURE 5.34 Position of Y-axis along paths 1–4. (From ref. [28].)

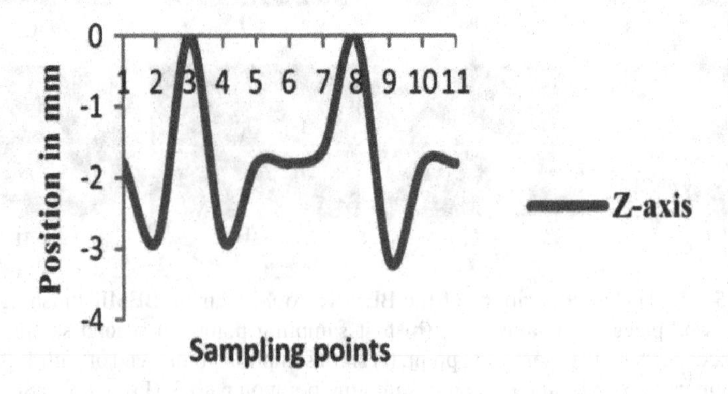

FIGURE 5.35 Position of Z-axis along paths 1–4. (From ref. [28].)

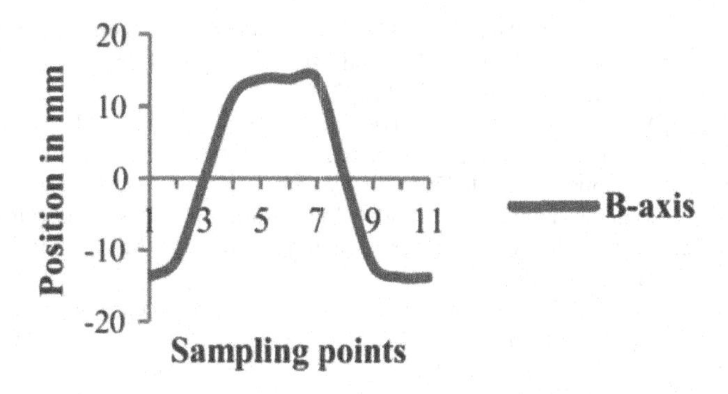

FIGURE 5.36 Position of B-axis along paths 1–4. (From ref. [28].)

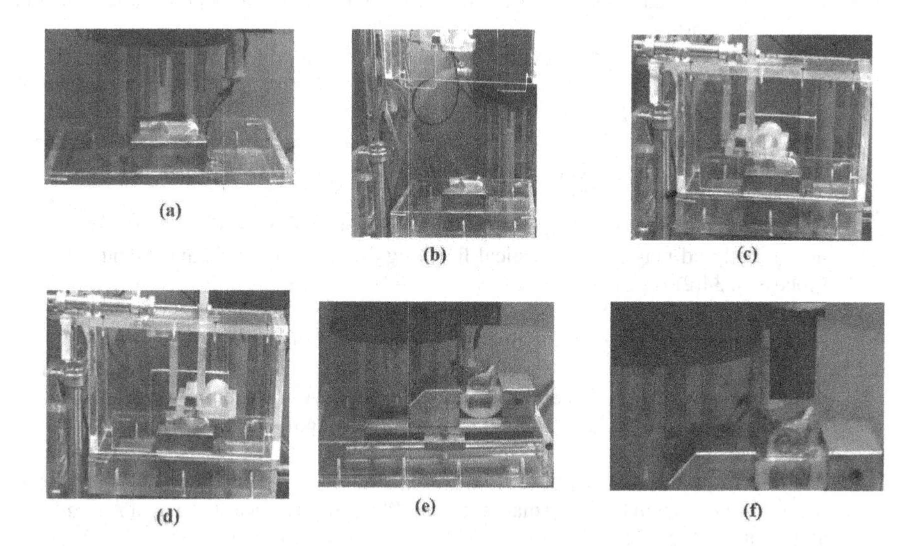

FIGURE 5.37 Different stages of integrated part-program of i5-B CNC BEMRF system. (a) Finishing cycle, (b) workpiece at cleaning cycle, (c) enclosure close for cleaning cycle, (d) cleaning cycle in progress, (e) workpiece at measurement cycle, (f) measurement cycle in progress. (From ref. [28].)

REFERENCES

1. Hung, C. L., Ku, W. L., & Yang, L. D. (2010). Prediction system of magnetic abrasive finishing (MAF) on the internal surface of a cylindrical tube. Materials and Manufacturing Processes, 25(12), 1404–1412.
2. Pandey, K., Pandey, U., & Pandey, P. M. (2019). Statistical modeling and surface texture study of polished silicon wafer Si (100) using chemically assisted double disk magnetic abrasive finishing. Silicon, 11(3), 1461–1479.
3. Khalaj Amineh, S., Fadaei Tehrani, A., Mosaddegh, P., & Mohammadi, A. (2015). A comprehensive experimental study on finishing aluminum tube by proposed UAMAF process. Materials and Manufacturing Processes, 30(1), 93–98.
4. Kordonski, W. I., & Jacobs, S. D. (1996). Magnetorheological finishing. International Journal of Modern Physics B, 10(23–24), 2837–2848.

5. Singh, A. K., Jha, S., & Pandey, P. M. (2011). Design and development of nanofinishing process for 3D surfaces using ball end MR finishing tool. International Journal of Machine Tools and Manufacture, 51(2), 142–151.
6. Kordonski, W. I., Shorey, A. B., & Tricard, M. (2004, January). Magnetorheological (MR) jet finishing technology. In ASME International Mechanical Engineering Congress and Exposition (Vol. 47098, pp. 77–84).
7. Sadiq, A., & Shunmugam, M. S. (2009). Magnetic field analysis and roughness prediction in magnetorheological abrasive honing (MRAH). Machining Science and Technology, 13(2), 246–268.
8. Jha, S., & Jain, V. (2006). Modeling and simulation of surface roughness in magnetorheological abrasive flow finishing (MRAFF) process. Wear, 261(7–8), 856–866.
9. Das, M., Jain, V. K., & Ghoshdastidar, P. S. (2011). The out-of-roundness of the internal surfaces of stainless-steel tubes finished by the rotational–magnetorheological abrasive flow finishing process. Materials and Manufacturing Processes, 26(8), 1073–1084.
10. Tiegelkamp, M., & John, K. H. (2010). IEC 61131-3: Programming Industrial Automation Systems. Springer.
11. Launhardt, M., Wörz, A., Loderer, A., Laumer, T., Drummer, D., Hausotte, T., & Schmidt, M. (2016). Detecting surface roughness on SLS parts with various measuring techniques. Polymer Testing, 53, 217–226.
12. Iqbal, F., Alam, Z., & Jha, S. (2020). Modelling of transient behaviour of roughness reduction in ball end magnetorheological finishing process. International Journal of Abrasive Technology, 10(3), 170–192.
13. Iqbal, F., & Jha, S. (2019). Experimental investigations into transient roughness reduction in ball-end magneto-rheological finishing process. Materials and Manufacturing Processes, 34(2), 224–231.
14. Iqbal, F., & Jha, S. (2018). Closed loop ball end magnetorheological finishing using in-situ roughness metrology. Experimental Techniques, 42(6), 659–669.
15. Alam, Z., Iqbal, F., Ganesan, S., & Jha, S. (2019). Nanofinishing of 3D surfaces by automated five-axis CNC ball end magnetorheological finishing machine using customized controller. The International Journal of Advanced Manufacturing Technology, 100(5), 1031–1042.
16. Alam, Z., Iqbal, F., & Jha, S. (2015). Automated control of three axis CNC ball end magneto-rheological finishing machine using PLC. International Journal of Automation and Control, 9(3), 201–210.
17. Iqbal, F., Alam, Z., Khan, D. A., & Jha, S. (2019). Constant work gap perpetuation in ball end magnetorheological finishing process. International Journal of Precision Technology, 8(2–4), 397–410.
18. Sahu, A. K., Iqbal, F., Kumar, A., & Jha, S. (2019). In situ geometric measurement of microchannels on EN31 steel by laser micromachining using confocal sensor. International Journal of Precision Technology, 8(2–4), 429–445.
19. Alam, Z., Khan, D. A., Iqbal, F., & Jha, S. (2019). Effect of polishing fluid composition on forces in ball end magnetorheological finishing process. International Journal of Precision Technology, 8(2–4), 365–378.
20. Khan, D. A., Alam, Z., Iqbal, F., & Jha, S. (2019). Experimental investigations on the effect of relative particle sizes of abrasive and iron powder in polishing fluid composition for ball end MR finishing of copper. International Journal of Precision Technology, 8(2–4), 354–364.
21. Alam, Z., & Jha, S. (2017). Modeling of surface roughness in ball end magnetorheological finishing (BEMRF) process. Wear, 374, 54–62.
22. Alam, Z., Khan, D. A., & Jha, S. (2018). A study on the effect of polishing fluid volume in ball end magnetorheological finishing process. Materials and Manufacturing Processes, 33(11), 1197–1204.

23. Alam, Z., Khan, D. A., & Jha, S. (2019). MR fluid-based novel finishing process for nonplanar copper mirrors. The International Journal of Advanced Manufacturing Technology, 101(1), 995–1006.
24. Kumar, A., Alam, Z., Khan, D. A., & Jha, S. (2019). Nanofinishing of FDM-fabricated components using ball end magnetorheological finishing process. Materials and Manufacturing Processes, 34(2), 232–242.
25. Khan, D. A., & Jha, S. (2018). Synthesis of polishing fluid and novel approach for nano-finishing of copper using ball-end magnetorheological finishing process. Materials and Manufacturing Processes, 33(11), 1150–1159.
26. Khan, D. A., & Jha, S. (2019). Selection of optimum polishing fluid composition for ball end magnetorheological finishing (BEMRF) of copper. The International Journal of Advanced Manufacturing Technology, 100(5), 1093–1103.
27. Iqbal, F., Rammohan, R., Patel, H. A., & Jha, S. (2016). Design and development of automated workpiece cleaning system for ball end magneto-rheological finishing process. In International Conference on Advances in Materials & Manufacturing 2016. ICAMM'16. International Conference on Dec 08 (pp. 289–295).
28. Iqbal, F., Alam, Z., Khan, D. A., & Jha, S. (2020). Part-Program-Based Process Control of Ball-End Magnetorheological Finishing. In Advances in Unconventional Machining and Composites (pp. 503–514). Springer, Singapore.

[illegible]

Index

K

L

M

N

O

P

R

S

T

U

V

Printed in the United States
by Baker & Taylor Publisher Services